Computational Intelligence for Engineering Systems

International Series on

INTELLIGENT SYSTEMS, CONTROL, AND AUTOMATION: SCIENCE AND ENGINEERING

VOLUME 46

For other titles published in this series, go to
www.springer.com/series/6259

Ana Madureira • Judite Ferreira • Zita Vale
Editors

Computational Intelligence for Engineering Systems

Emergent Applications

Editors
Ana Madureira
Computer Science Department
School of Engineering-Polytechnic of Porto
Porto-Portugal
amd@isep.ipp.pt

Judite Ferreira
Electrical Engineering Department
School of Engineering-Polytechnic of Porto
Porto-Portugal
mju@isep.ipp.pt

Zita Vale
Electrical Engineering Department
School of Engineering-Polytechnic of Porto
Porto-Portugal
zav@isep.ipp.pt

ISBN 978-94-007-3444-9 ISBN 978-94-007-0093-2 (eBook)
DOI 10.1007/978-94-007-0093-2
Springer Dordrecht Heidelberg London New York

Cover design: Spi Publisher Services

Printed on acid-free paper

Springer is part of Springer Science+Business Media (www.springer.com)

Preface

Computational Intelligence can be seen as a science, as it seeks, studies and tries to understand the phenomenon of intelligence, and as a branch of engineering, as it seeks to build tools to assist and support human intelligence. Providing computers with intelligence that might be useful to human activity is the major goal of Computational Intelligence research projects.

The complexity of current computer systems has led software engineering, distributed systems and management communities to look for inspiration in diverse fields, such as robotics, artificial intelligence or biology, in order to find new ways of designing and managing systems. Looking at processes that can be found in nature, it is possible to try to understand and mimic them to solve complex problems on different domains.

This book addresses, in a single volume, contributions in Emergent Applications of Computational Intelligence for Engineering Systems, selected from the works presented at the International Symposium on Computational Intelligence for Engineering Systems (ISCIES'09)) held in the School of Engineering of the Polytechnic of Porto, Portugal, November 19-20, 2009.

ISCIES'09 provided a forum to discuss the state-of-the-art, recent research results and perspectives of future developments with respect to the symposium themes. ISCIES'09 provided a stimulating discussion for scientists, engineers, educators, and students to disseminate the latest research results and exchange information on emerging areas of research in the field of Computational Intelligence. ISCIES'09 also aimed at identifying new Computational Intelligence technologies and emergent areas for intelligent systems applications.

Sensors and Smart Services, Decision Support Systems, Ambient Intelligence, Intelligent Energy Systems, Intelligent Manufacturing Systems, Intelligent Systems Inspired by Nature, Computational Creativity, Autonomous Mental Development, Bioinformatics, Bioengineering and Autonomic Computing are some of the themes that are addressed in the present volume.

We would like to thank all referees and other colleagues who helped in the edition process of this book. Our thanks are also due to all participants for their contributions to the ISCIES'09 Symposium and to this book.

Finally, the editors would like to acknowledge FCT (Portuguese Science and Technology Foundation) for its support to GECAD - Knowledge Engineering and Decision Support Group Unit activities and initiatives.

Ana Madureira
Computer Science Department
School of Engineering-Polytechnic of Porto

Judite Ferreira
Electrical Engineering Department
School of Engineering-Polytechnic of Porto

Zita Vale
Electrical Engineering Department
School of Engineering-Polytechnic of Porto

Contents

Intention Recognition with Evolution Prospection and Causal Bayes Networks

Luís Moniz Pereira and Han The Anh

Centro de Inteligência Artificial (CENTRIA) Departamento de Informática, Faculdade de Ciências e Tecnologia,
Universidade Nova de Lisboa, 2829-516 Caparica, Portugal
(email: {lmp, h.anh}@fct.unl.pt)

Abstract We describe a novel approach to tackle intention recognition, by combining dynamically configurable and situation-sensitive Causal Bayes Networks plus plan generation techniques. Given some situation, such networks enable the recognizing agent to come up with the most likely intentions of the intending agent, i.e. solve one main issue of intention recognition. And, in case of having to make a quick decision, focus on the most important ones. Furthermore, the combination with plan generation provides a significant method to guide the recognition process with respect to hidden actions and unobservable effects, in order to confirm or disconfirm likely intentions. The absence of this articulation is a main drawback of the approaches using Bayes Networks solely, due to the combinatorial problem they encounter. We explore and exemplify its application, in the Elder Care context, of the ability to perform Intention Recognition and of wielding Evolution Prospection methods to help the Elder achieve its intentions. This is achieved by means of an articulate use of a Causal Bayes Network to heuristically gauge probable general intention – combined with specific generation of plans involving preferences – for checking which such intentions are plausibly being carried out in the specific situation at hand, and suggesting actions to the Elder. The overall approach is formulated within one coherent and general logic programming framework and implemented system. The paper recaps required background and illustrates the approach via an extended application example.

Keywords: Intention recognition, Elder Care, Causal Bayes Networks, Plan generation, Evolution Prospection, Preferences, Logic Programming.

A. Madureira et al. (eds.), *Computational Intelligence for Engineering Systems: Emergent Applications*, Intelligent Systems, Control and Automation: Science and Engineering 46,
DOI 10.1007/978-94-007-0093-2_1, © Springer Science + Business Media B.V. 2011

1 Introduction

In many multi-agent systems, the problem of intention recognition appears to be crucial when the agents cooperate or compete to achieve a certain task, especially when the possibility of communication is limited. For example, in heterogeneous agent systems it is likely that agents speak different languages, have different designs or different levels of intelligence; hence, intention recognition may be the only way the agents understand each other so as to secure a successful cooperation. Moreover, when competing, the agents even often attempt to hide their real intentions and make others believe in some pretense ones. Intention recognition in this setting becomes undoubtedly crucial for agents, in order to prepare themselves from potential hostile behaviors from others.

Needless to say, the recognized intentions provide the recognizing agent with valuable information in dealing with other agents, whether they cooperate or compete with each other. But how this information can be valuable for the recognizing agent? In this work, besides the problem of intention recognition, we attempt to address that issue using our implemented Evolution Prospection Agent system [Pereira and Anh 2009b, Pereira and Anh 2009c].

Recently, there have been many works addressing the problem of intention recognition as well as its applications in a variety of fields. As in Heinze's doctoral thesis [Heinze 2003], intention recognition is defined, in general terms, as the process of becoming aware of the intention of another agent and, more technically, as the problem of inferring an agent's intention through its actions and their effects on the environment. According to this definition, one approach to tackle intention recognition is by reducing it to plan recognition, i.e. the problem of generating plans achieving the intentions and choosing the ones that match the observed actions and their effects in the environment of the intending agent. This has been the main stream so far [Heinze 2003, Kautz and Allen 1986].

One of the core issues of that approach is that of finding an initial set of possible intentions (of the intending agent) that the plan generator is going to tackle, and which must be imagined by the recognizing agent. Undoubtedly, this set should depend on the situation at hand, since generating plans for all intentions one agent could have, for whatever situation he might be in, is unrealistic if not impossible.

In this work, we use an approach to solve this problem employing so-called situation-sensitive *Causal Bayes Networks* (CBN) - that is, CBNs [Glymor 2001] that change according to the situation under consideration, itself subject to ongoing change as a result of actions. Therefore, in some given situation, a CBN can be configured dynamically, to compute the likelihood of intentions and filter out the much less likely ones. The plan generator (or plan library) thus only needs, at the start, to deal with the remaining more relevant because more probable or credible intentions, rather than all conceivable intentions. One of the important advantages of our approach is that, on the basis of the information provided by the CBN the recognizing agent can see which intentions are more likely and worth addressing, so, in case of having to make a quick decision, it can focus on the most relevant ones first. CBNs, in our work, are represented in P-log [Baral et al. 2004, Anh et al. 2009, Baral et al. 2009], a declarative language that combines logical and probabilistic reasoning, and uses Answer Set Programming (ASP) as it's logical and CBNs as its probabilistic foundations. Given a CBN, its situation-sensitive version is constructed by attaching to it a logical component to dynamically compute situation specific probabilistic information, which is forthwith inserted into the P-log program representing that CBN. The computation is dynamic in the sense that there is a process of inter-feedback between the logical component and the CBN, i.e. the result from the updated CBN is also given back to the logical component, and that might give rise to further updating, etc.

In addition, one more advantage of our approach, in comparison with the stream of those using solely BNs [Tahbounb 2006, Schrempf et al. 2007] is that these just use the available information for constructing CBNs. For complicated tasks, e.g. in recognizing hidden intentions, not all information is observable. Whereas CBNs are appropriate for coding general average information, they quickly bog down in detail when aspiring to code multitudes of specific situations and their conditional probability distributions. The approach of combining CBNs with plan generation provides a way to guide the recognition process: which actions (or their effects) should be checked whether they were (hiddenly) executed by the intending agent. So, plan recognition ties the average statistical information with the situation particulars, and obtains specific situational information that can be fed into the CBN. In practice, one can make use of any plan generators or plan libraries available. For integration's sake, we can use the ASP based conditional planner called ASCP [Tu et Al. 2007] from XSB Prolog using the XASP package [Castro et al., The XSB System 2000] for interfacing with Smodels [Nimelä and Simons 1997] an answer set solver or, alternatively, rely on plan libraries so obtained.

The next step, which of taking advantage of the recognized intention gleaned from the previous stage, is implemented using our Evolution Prospection Agent (EPA) system [Pereira and Anh 2009b, Pereira and Anh 2009c]. The latter allows an agent to be able to look ahead, prospectively, into its hypothetical futures, in order to determine the best courses of evolution that satisfy its goals, and thence to prefer amongst those futures. These courses of evolution can be provided to the intending agent as suggestions to achieve its intention (in cooperating settings) or else as a guide to prevent that agent from achieving it (in hostile settings).

In EPA system, *a priori* and *a posteriori* preferences, embedded in the knowledge representation theory, are used for preferring amongst hypothetical futures. The *a priori* ones are employed to produce the most interesting or relevant conjectures about possible future states, while the *a posteriori* ones allow the agent to actually make a choice based on the imagined consequences in each scenario. In addition, different kinds of evolution-level preferences enable agents to attempt long-term goals, based on the historical information as well as quantitative and qualitative *a posteriori* evaluation of the possible evolutions.

In the sequel we describe the intention recognition and evolution prospection systems, showing an extended example for illustration. Then, Elder Care – a real world application domain is addressed by the combination of the two systems. The paper finishes with Conclusions and Future directions.

2 Intention Recognition

2.1 Causal Bayes Networks

We briefly recall Causal Bayes Networks (CBN) here for convenience in order to help understand their use for intention recognition and their realization in P-log. Firstly, let us recall some preliminary definitions.

Definition 1 (Directed Acyclic Graph). A directed acyclic graph, also called a dag, is a directed graph with no directed cycles; that is, for any node $\boldsymbol{v}$, there is no non-empty directed path that starts and ends on $\boldsymbol{v}$.

Definition 2. Let $\boldsymbol{G}$ be a dag that represents causal relations between its nodes. For two nodes $\boldsymbol{A}$ and $\boldsymbol{B}$ of $\boldsymbol{G}$, if there is an edge from $\boldsymbol{A}$ to $\boldsymbol{B}$ (i.e. $\boldsymbol{A}$ is a direct cause of $\boldsymbol{B}$), $\boldsymbol{A}$ is called a parent of $\boldsymbol{B}$, and $\boldsymbol{B}$ is a child of $\boldsymbol{A}$. The set of parent nodes of a node $\boldsymbol{A}$ is denoted by $\boldsymbol{parents}\ (\boldsymbol{A})$. Ancestor nodes of $\boldsymbol{A}$ are parents of $\boldsymbol{A}$ or parents of some ancestor nodes of A. If node A has no parents ($\boldsymbol{parents}(\boldsymbol{A}) = \boldsymbol{0}$), it is called a ***top node***. If A has no child, it is called a ***bottom node***. The nodes which are neither top nor bottom are said ***intermediate***. If the value of a node is observed, the node is said to be an ***evidence node***.

Definition 3 (Causally Sufficient Set [Glymour 2001]*).* Let $\boldsymbol{V}$ be a set of variables with causal relations represented by a dag. $\boldsymbol{V}$ is said to be causally sufficient if and only if for any $\boldsymbol{Y}, \boldsymbol{Z} \in \boldsymbol{V}$ with $\boldsymbol{Y} \neq \boldsymbol{Z}$ and $\boldsymbol{X}$ is a common cause of $\boldsymbol{Y}$ and $\boldsymbol{Z}$, then $\boldsymbol{X} \in \boldsymbol{V}$. That is to say, $\boldsymbol{V}$ is causally sufficient if for any two distinct variables in $\boldsymbol{V}$, all their common causes also belong to $\boldsymbol{V}$.

Definition 4 (Causal Markov Assumption - CMA [Glymour 2001]*).* Let $\boldsymbol{X}$ be any variable in a causally sufficient set $\boldsymbol{S}$ of variables or features whose causal relations are represented by a dag $\boldsymbol{C}$, and let $\boldsymbol{P}$ be the set of all variables in $\boldsymbol{S}$ that are direct causes of $\boldsymbol{X}$ (i.e. parents of $\boldsymbol{X}$ in $\boldsymbol{G}$). Let $\boldsymbol{Y}$ be any subset of $\boldsymbol{S}$ such that no variable in $\boldsymbol{Y}$ is a direct or indirect effect of $\boldsymbol{X}$ (i.e., there is no directed path in $\boldsymbol{G}$ from $\boldsymbol{X}$ to any member of $\boldsymbol{Y}$). Then $\boldsymbol{X}$ is independent (in probability) of $\boldsymbol{Y}$ conditional on $\boldsymbol{P}$.

The CMA implies that the joint probability of any set of values of a causally sufficient set can be factored into a product of conditional probabilities of the value of each variable on its parents. More details and a number of examples can be found in [Gylmour 2001].

Definition 5 (Bayes Networks). A Bayes Network is a pair consisting of a dag whose nodes represent variables and missing edges encode conditional independencies between the variables, and an associated probability distribution satisfying the Causal Markov Assumption.
If the dag of a BN is intended to represent causal relations and its associated probability distribution is intended to represent those that result from the represented mechanism, then the BN is said to be causal. To do so, besides CMA, the associated probability distribution needs to satisfy an additional condition, as more formally shown in the following definition.

Definition 6 (Causal Bayes Network). A Bayes Network is causal if its associated probability distribution satisfies the condition specifying that if a node $\boldsymbol{X}$ of its dag is actively caused to be in a given state $\boldsymbol{x}$ (an operation written as $\boldsymbol{do(x)}$, e.g. in P-log syntax), then the probability density function changes to the one of the network obtained by cutting the links from $\boldsymbol{X}$'s parents to $\boldsymbol{X}$, and setting $\boldsymbol{X}$ to the caused value $\boldsymbol{x}$ [Pearl 2000].

With this condition being satisfied, one can predict the impact of external interventions from data obtained prior to intervention.
In a BN, associated with each intermediate node of its dag is a specification of the distribution of its variable, say A, conditioned on its parents in the graph, i.e. $P(A|parents(A))$ is specified. For a top node, the unconditional distribution of the variable is specified. These distributions are called Conditional Probability Distribution (CPD) of the BN.
Suppose nodes of the dag form a causally sufficient set, i.e. no common causes of any two nodes are omitted, then implied by CMA [Gylmour 2003], the joint distribution of all node values of a causally sufficient can be determined as the product of conditional probabilities of the value of each node on its parents

$$P(X_1, \dots X_N) = \prod_{i=1}^{N} P(X_i|parents(X_i))$$

where $V = \{Xi|1 \leq i \leq N\}$ is the set of nodes of the dag.
Suppose there is a set of evidence nodes in the dag, say $O = \{O1, \dots, Om\} \subset V$. We can determine the conditional probability of a variable X given the observed value of evidence nodes by using the conditional probability formula

$$P(X|O) = \frac{P(X,O)}{P(O)} = \frac{P(X, O_1, \dots, O_n)}{P(O_1, \dots, O_n)} \quad (1)$$

where the numerator and denominator are computed by summing the joint probabilities over all absent variables w.r.t. V as follows

$$P(X = x, O = o) = \sum_{av \in ASG(AV_1)} P(X = x, O = o, AV_1 = av)$$

$$P(O = o) = \sum_{av \in ASG(AV_2)} P(O = o, AV_2 = av)$$

where $o = \{o_1, \ldots, o_m\}$ with $o_1, \ldots, o_m$ being the observed values of $O_1, \ldots, O_m$, respectively; ASG(Vt) denotes the set of all assignments of vector Vt (with components are variables in V); AV_1, AV_2 are vectors components of which are corresponding absent variables, i.e. variables in $V \setminus \{O \cup \{X\}\}$ and $V \setminus O$, respectively.

In short, to define a BN, one needs to specify the structure of the network, its CPD and the prior probability distribution of the top nodes. We will see some examples later, in Fig. 1.3 and Fig. 1.9.

2.2 Intention recognition with Causal Bayesian Networks

The first phase of the intention recognition system is to find out how likely each conceivable intention is, based on current observations such as observed actions of the intending agent or the effects its actions (either actually observed or missed direct observation) have in the environment. A conceivable intention is the one having causal relations to all current observations. It is carried out by using a CBN with nodes standing for binary random variables that represent causes, intentions, actions and effects. The structure of the CBN and its components such as CPD and probability distribution of top nodes are specified in the sequel.

To begin with, for better understand the structure of CBNs for intention recognition we will present, let us recall the high-level model of intentional behavior described in [Heinze 2003]. In this work, Heinze [Heinze 2003] proposed a tri-level decompositional model of intentional behavior of the intending agent. Intention recognition is the reversal of this process. These levels are:

- *The intentional level* describes the intentions of agent in terms of desires, beliefs, goals, plans and other high-level intentional states. These intentions give rise, in a directly causal way, to activities.
- *The activity level* describes the activities and actions undertaken by the agent. The actions are a direct result of the intentional state of the agent. The activities can be to select plans, adopt tactics, etc.
- *The state level* describes the agent in terms of externally accessible characteristics and effects its actions have in the environment.

According to this description, six basic paths (approaches) can be followed in intention recognition as shown in Fig. 1.1. For example, scheme 1 corresponds to the basic *'sense and infer'* approach, scheme 3 matches the trivial case of communicating intentions, while scheme 6 resembles direct action recognition from state and inferring the recognized action in intention.

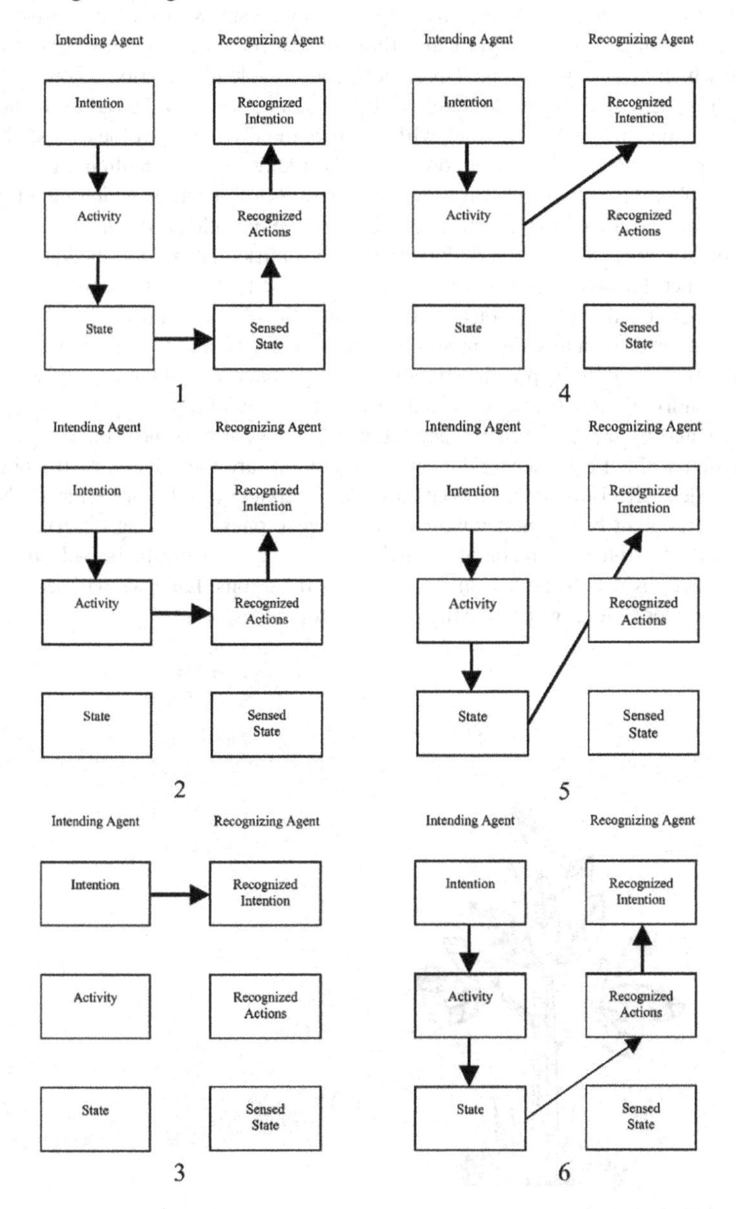

Fig. 1.1 Six Possible Paths in intention Recognition [Heinze 2003]

We may also think of these six basic paths as all possible cases that can happen when recognizing intentions of an agent, depending on the information available. For example, when not able to observe actions of the intending agent, the recognizing agent must attempt to analyze the accessible states of the environment to figure out which actions might cause those states to the environment. Depending on to which level of intentional behavior the observable information belongs to, an intention recognition system should be able to flexibly employ the right scheme.

Based on this tri-level model with a small modification, we next describe a structure of CBNs that allows computing the likelihood of intentions, given the observable information. A fourth level that describes the causes which might give rise to the considered intentions is added. That level is called *pre-intentional.*

The structure is as Fig. 1.1. Intentions are represented by intermediate nodes whose ancestor nodes represent causes that give rise to those intentions. The causes, as mentioned above, belong to the pre-intentional level of a model whose intentional level contains the intentions. Intuitively, the additional level is introduced, first of all, to help with estimating prior probabilities of the intentions.

Secondly, it guarantees the causal sufficiency condition of the set of variables represented by the nodes of the dag. However, if these prior probabilities can be specified without considering the causes, intentions are represented by top nodes. Top nodes reflect the problem context or the intending agent's mental state. Note that there might be top nodes which are evidence ones, i.e. being observable. In our CBN for intention recognition, evidence nodes need not to be only the observed actions which result from having some intentions. Later we will see an example (Elder Care, Fig. 1.9) having observed top nodes.

Fig. 1.2 Fox and Crow

Observed actions are represented as children of the intentions that causally affect them. Observable effects are represented as bottom nodes. They can be children of observed action nodes, of intention nodes, or of some unobserved actions that might cause the observable effects that are added as children of the intention nodes.

The above causal relations (e.g. which causes give rise to an intention, which intentions trigger an action, which actions have an effect) among nodes of the BNs, as well as its CPD and the distribution of the top nodes, are specified by domain experts. However, they are also possible to learn automatically. Finally, by using formula (1) the conditional probabilities of each intention on current observations can be determined, X being an intention and O being the set of current observations.

Example 1.1 (Fox-Crow). Consider Fox-Crow story - adapted from Aesop's fable (Fig. 1.2). There is a crow, holding a cheese. A fox, being hungry, approaches the crow and praises her, hoping that the crow will sing and the cheese will fall down near him. Unfortunately for the fox, the crow is very intelligent, having the ability of intention recognition.

The Fox's intentions CBN is depicted in the Fig. 1.3. The initial possible intentions of Fox that Crow comes up with are: Food - $i(F)$, Please - $i(P)$ and Territory - $i(T)$. The facts that might give rise to those intentions are how friendly the Fox is ($Friendly\_fox$) and how hungry he is ($Hungry\_fox$). Currently, there is only one observation which is: Fox praised Crow ($Praised$).

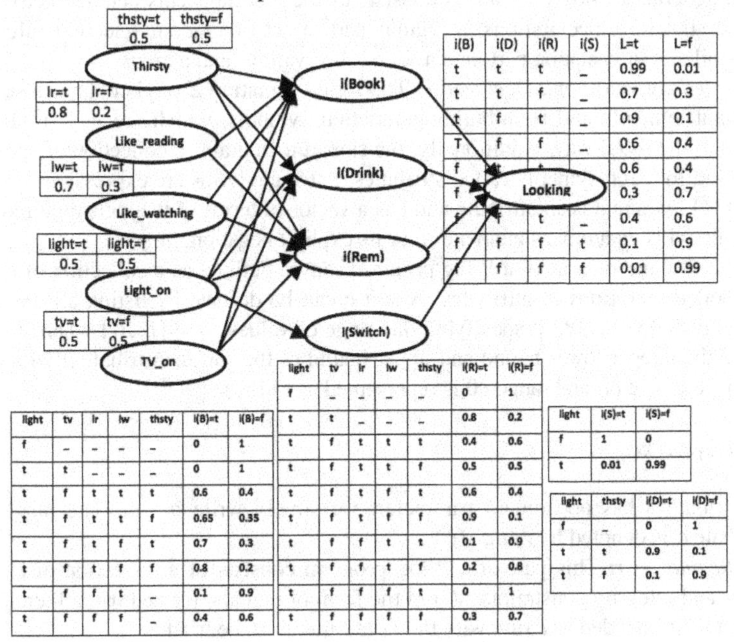

thsty=t	thsty=f
0.5	0.5

lr=t	lr=f
0.8	0.2

lw=t	lw=f
0.7	0.3

light=t	light=f
0.5	0.5

tv=t	tv=f
0.5	0.5

i(B)	i(D)	i(R)	i(S)	L=t	L=f
t	t	t	_	0.99	0.01
t	t	f	_	0.7	0.3
t	f	t	_	0.9	0.1
T	f	f	_	0.6	0.4
f	t	t	[illegible]	0.6	0.4
f	[illegible]	[illegible]	[illegible]	0.3	0.7
f	f	t	_	0.4	0.6
f	f	f	t	0.1	0.9
f	f	f	f	0.01	0.99

light	tv	lr	lw	thsty	I(B)=t	I(B)=f
f	_	_	_	_	0	1
t	t	_	_	_	0	1
t	f	t	t	t	0.6	0.4
t	f	t	t	f	0.65	0.35
t	f	t	f	t	0.7	0.3
t	f	t	f	f	0.8	0.2
t	f	f	t	_	0.1	0.9
t	f	f	f	_	0.4	0.6

light	tv	lr	lw	thsty	I(R)=t	I(R)=f
f	_	_	_	_	0	1
t	t	_	_	_	0.8	0.2
t	f	t	t	t	0.4	0.6
t	f	t	t	f	0.5	0.5
t	f	t	f	t	0.6	0.4
t	f	t	f	f	0.9	0.1
t	f	f	t	t	0.1	0.9
t	f	f	t	f	0.2	0.8
t	f	f	f	t	0	1
t	f	f	f	f	0.3	0.7

light	I(S)=t	I(S)=f
f	1	0
t	0.01	0.99

light	thsty	I(D)=t	I(D)=f
f	_	0	1
t	t	0.9	0.1
t	f	0.1	0.9

Fig. 1.3 Fox's Intentions CBN

2.3 P-log

The computation in CBNs is automated using P-log, a declarative language that combines logical and probabilistic reasoning, and ASP as it's logical and CBNs as its probabilistic foundations. We recap it here for self-containment, to the extent we use it.

The original P-log [Baral et al. 2004, Baral et al. 2009] uses ASP as a tool for computing all stable models of the logical part of P-log. Although ASP has been proved to be a useful paradigm for solving a variety of combinatorial problems, its non-relevance property [Castro et al.] makes the P-log system sometimes computationally redundant. Newer developments of P-log [Castro et al.] use the XASP package of XSB Prolog [The XSB System 2000] for interfacing with Smodels [The XSB System 2000] – an answer set solver. The power of ASP allows the representation of both classical and default negation in P-log easily. Moreover, the new P-log uses XSB as the underlying processing platform, allowing arbitrary Prolog code for recursive definitions. Consequently, it allows more expressive queries not supported in the original version, such as meta queries (probabilistic built- in predicates can be used as usual XSB predicates, thus allowing full power of probabilistic reasoning in XSB) and queries in the form of any XSB predicate expression [Anh et al. 2008]. Moreover, the tabling mechanism of XSB [Swift 1999] significantly improves the performance of the system.

In general, a P-log program Π consists of the 5 components detailed below: a sorted signature, declarations, a regular part, a set of random selection rules, a probabilistic information part, and a set of observations and actions.

(i) Sorted signature and Declaration The sorted signature Σ of Π contains a set of constant symbols and term-building function symbols, which are used to form terms in the usual way. Additionally, the signature contains a collection of special reserved function symbols called attributes. Attribute terms are expressions of the form $(\bar{t})$, where a is an attribute and $\bar{t}$ is a vector of terms of the sorts required by a. A literal is an atomic statement, a, or its explicit negation, neg_p.

The declaration part of a P-log program can be defined as a collection of sorts and sort declarations of attributes. A sort c can be defined by listing all the elements $c = \{x_1, \ldots, x_n\}$, specifying the range of values $c = \{L..U\}$ where L and U are the integer lower bound and upper bound of the sort c. Attribute a with domain $c1\ x \ldots x\ cn$ and range $c0$ is represented as follows:

$$a: c_1 x \ldots x c_n \ --> \ c_0$$

If attribute a has no domain parameter, we simply write $a : c_0$. The range of attribute a is denoted by $range(a)$.

(ii) Regular part This part of a P-log program consists of a collection of rules, facts, and integrity constraints (IC) in the form of denials, formed using literals of Σ. An IC is encoded as a rule with the $false$ literal in the head.

(iii) Random Selection Rule This is a rule for attribute a having the form:

$$random(RandomName, a(\bar{t}), DynamicRange) :- Body$$

This means that the attribute instance $a(\bar{t})$ is random if the conditions in $Body$ are satisfied. The $DynamicRange$, not used in the particular examples in the sequel, allows to restrict the default range for random attributes. The $RandomName$ is a syntactic mechanism used to link random attributes to the corresponding probabilities. If there is no precondition, we simply put true in the body. A constant $full$ can be used in $DynamicRange$ to signal that the dynamic domain is equal to $range(a)$.
(iv) Probabilistic Information. Information about probabilities of random attribute instances $a(\bar{t})$ taking a particular value y is given by probability atoms (or simply pa-atoms) which have the following form:

$$pa\left(RandomName, a(\bar{t}, y), d\ (A, B)\right) :- Body.$$

meaning that if the $Body$ were true, and the value of $a(\bar{t})$ were selected by a rule named $RandomName$, then $Body$ would cause $a(\bar{t}) = y$ with probability $\frac{A}{B}$.
(v) Observations and Actions These are, respectively, statements of the forms $obs(l)$ and $do(l)$, where l is a literal. Observations are used to record the outcomes of random events, i.e. of random attributes and attributes dependent on them. The statement $do(a(t, y))$ indicates that $a(t) = y$ is enforced true as the result of a deliberate action, not an observation.

2.4 Recognizing Fox's intentions - An Example

Example 1.2 (Fox-Crow – Example 1.1 cont'd). The Fox's intentions CBN can be coded with the P-log program in Fig. 1.4.

Two sorts $bool$ and $fox\_intentions$, in order to represent boolean values and set of Fox's intentions, are declared in part 1. Part 2 is the declaration of four attributes $hungry\_fox$, $friendly\_fox$, $praised$ and i which state the first three attributes have no domain parameter and get boolean values, and the last one maps each Fox's intention to a boolean value. The random selection rules in part 3 declare that these four attributes are randomly distributed in their ranges. The distributions of the top nodes ($hungry\_fox$, $friendly\_fox$) and the CPD corresponding to the CBN in Fig. 1.3 are given in part 4 and parts 5-8, respectively, using the probabilistic information pa-rules. For example, in part 4 the first rule says that fox is hungry with probability 1/2 and the second rule says he is friendly with probability 1/100. The first rule in part 5 states that if Fox is friendly and hungry, the probability of him having intention Food is 8/10.

Note that the probability of an atom $a(\bar{t}, y)$ will be directly assigned if the corresponding *pa*/3 atom is in the head of some *pa-rule* with a true body. To define probabilities of the remaining atoms we assume that by default, all values of a given attribute which are not assigned a probability are equally likely. For example, first rule in part 4 implies that fox is not hungry with probability 1/2. And, actually, we can remove that rule without changing the probabilistic information since, in that case, the probability of fox being hungry and of not being so are both defined by default, thus, equal to 1/2.

The probabilities of Fox having intention Food, Territory and Please given the observation that Fox praised Crow can be found in P-log with following queries, respectively,

$? - pr(i(food,t) \mid obs(praised(t)), V1).$ $The\ answer\ is: V1 = 0.9317.$
$? - pr(i(territory,t) \mid obs(praised(t)), V2).$ $The\ answer\ is: V2 = 0.8836.$
$? - pr(i(please,t) \mid obs(praised(t)), V3).$ $The\ answer\ is: V3 = 0.0900.$

```
1. bool = {t,f}.          fox_intentions = {food,please,territory}.
2. hungry_fox : bool.                        friendly_fox : bool.
   i : fox_intentions --> bool.              praised : bool.
3. random(rh, hungry_fox, full).  random(rf, friendly_fox, full).
   random(ri, i(I), full).        random(rp, praised, full).
4. pa(rh,hungry_fox(t),d_(1,2)).  pa(rf,friendly_fox(t),d_(1,100)).
5. pa(ri(food),i(food,t),d_(8,10)) :-friendly_fox(t),hungry_fox(t).
   pa(ri(food),i(food,t),d_(9, 10)) :-friendly_fox(f) ,hungry_fox(t).
   pa(ri(food),i(food,t),d_(0.1,10)):-friendly_fox(t),hungry_fox(f).
   pa(ri(food),i(food,t),d_(2, 10)) :-friendly_fox(f) ,hungry_fox(f).
6. pa(ri(please),i(please,t),d_(7,10))        :-friendly_fox(t),hungry_fox(t).
   pa(ri(please),i(please,t) ,d_(1,100))      :-friendly_fox(f),hungry_fox(t).
   pa(ri(please),i(please,t) ,d_(95, 100))    :-friendly_fox(t),hungry_fox(f).
   pa(ri(please),i(please,t) ,d_(5,100))      :-friendly_fox(f),hungry_fox(f).
7. pa(ri(territory),i(territory,t),d_(1,10))  :-friendly_fox(t).
   pa(ri(territory),i(territory,t),d_(9, 10)) :-friendly_fox(f).
8. pa(rp, praised(t),d_(95,100))  :-i(food,t),i(please,t).
   pa(rp, praised(t),d_(6,10))    :-i(food,t),i(please,f).
   pa(rp, praised(t),d_(8,10))    :-i(food,f),i(please,t).
   pa(rp, praised(t),d_(1,100))   :-i(food,f),i(please,f),i(territory,t).
   pa(rp, praised(t),d_(1,1000))  :-i(food,f),i(please,f),i(territory,f)
```

Fig. 1.4 Fox's intentions CBN

From the result we can say that Fox is very unlikely to have the intention Please, i.e. to make the Crow pleased since its likelihood is very much less than the others. Thus, the next step of Crow's intention recognition is to generate conceivable plans that might corroborate the two remaining intentions. The one with greater likelihood will be discovered first.

2.5 Situation-sensitive CBNs

Undoubtedly, CBNs should be situation-sensitive since using a general CBN for all specific situations (instances) of a problem domain is unrealistic and most likely imprecise. For example, in the Fox-Crow scenario the probabilistic information in Crow's CBN about the Fox's intention of getting Crow's territory very much depends on what kind of territories the Crow occupies. However, consulting the domain expert to manually change the CBN w.r.t. each situation is also very costly. We here provide a way to construct situation-sensitive CBNs, i.e. ones that change according to the given situation. It uses Logic Programming (LP) techniques to compute situation specific probabilistic information which is then updated into a CBN general for the problem domain.

The LP techniques can be deduction with top-down procedure (Prolog) (to deduce situation-specific probabilistic information) or abduction (to abduce probabilistic information needed to explain observations representing the given situation). However, we do not exclude various other types of reasoning, e.g. including integrity constraint satisfaction, abduction, contradiction removal, preferences, or inductive learning, whose results can be compiled (in part) into an evolving CBN.

The issue of how to update a CBN with new probabilistic information can take advantage of the advance in LP semantics for evolving programs with updates [Alferes et al. 2002, Alferes et al. 2005, Alferes et al. 2000]. However, in this work we employ a simpler way, demonstrated in the following example.

Example 1.3 (Fox-Crow (cont'd)). Suppose the fixed general CBN is the one given in Fig. 1.4. The Prolog program contains the following two rules for updating the probabilistic information in part 7 of the CBN:

```
pa_rule(pa(ri(territory),i(territory,t) ,d_(0,100)), [friendly_fox(t)])
                        :- territory(tree).
pa_rule(pa(ri(territory),i(territory,t) ,d_(1,100)), [friendly_fox(f)])
                        :- territory(tree).
```

Given a P-log probabilistic information pa-rule, then the corresponding so-called situation-sensitive $pa\_rule/2$ predicate takes the head and body of the pa-rule as its first and second arguments, respectively. A situation is given, in this work, by asserted facts representing it. In order to find the probabilistic information specific for the given situation, we simply use the XSB built-in $findall/3$ predicate to find all true $pa\_rule/2$ literals.

In the story the Crow's territory is a tree, thus the fact $territory(tree)$ is asserted. Hence, the following two $pa\_rule/2$ literals are true

```
pa_rule(pa(ri(territory),i(territory,t) ,d_(0,100)), [friendly_fox(t)])
pa_rule(pa(ri(territory),i(territory,t) ,d_(1,100)), [friendly_fox(f)])
```

The CBN is updated by replacing the two pa-rules in part 7 of the CBN with the corresponding two rules

```
pa(ri(territory) ,i(territory,t) ,d_(0, 100)) :- friendly_fox(t)
pa(ri(territory) ,i(territory,t) ,d_(1, 100)) :- friendly_fox(f)
```

This change can be easily made at the preprocessing stage of the implementation of P-log(XSB) (more details about the system implementation can be found in [Anh et al. 2008]).

In this updated CBN the likelihood of the intentions $i(food, t)$, $i(territory, t)$, $i(please, t)$ are: $V1 = 0.9407$; $V2 = 0.0099$; $V3 = 0.0908$, respectively. Thus, more likely, the only surviving intention is $food$.

2.6 Plan Generation

The second phase of the intention recognition system is to generate conceivable plans that can achieve the most likely intentions surviving after the first phase. Any appropriate planners, though those implemented in ASP and/or Prolog are preferable for integration's sake, might be used for this task, e.g. DLV^K– a declarative, logic-based planning system built on top of the DLV [Eiter et al. 2003] and $ASCP$ – an ASP based conditional planner [Tu et al. 2007].

In our system, plan generation is carried out by a new implementation of $ASCP$ in XSB Prolog using $XASP$ package [An implementation of ASCP using XASP]. It has the same syntax and uses the same transformation to ASP as in the original version. It might have better performance because of the relevance property and tabling mechanism in XSB, but we will not discuss that here. Next we briefly recall the syntax of $ASCP$ necessary to represent the example being considered. Semantics and the transformation to ASP can be found in [Tu et al. 2007].

2.7 Action language A_k^c

ASCP uses A_k^c - a representation action language that extends A[Gelfond and Lifschitz 1993] by introducing new types of propositions called *knowledge producing proposition and executability condition*, and *static causal laws*.

The alphabet of A_k^c consists of a set of actions A and a set of fluents F. A *fluent literal* (or *literal* for short) is either a fluent $f \in F$ or its negation $-f$. A fluent formula is a propositional formula φ constructed from the set of literals using operators $\wedge$, V and/or $\neg$. To describe an action theory, 5 kinds of propositions used: (1) $initially(l)$; (2) $executable(a, \psi)$; (3) $causes(a, l, \varphi)$; (4) $if(l, \phi)$; and (5) $determines(a, 9)$.

The initial situation is described by a set of propositions (1), called v- propositions. (1) says that l holds in the initial situation. A proposition of form (2) is called executability condition. It says that a is executable in any situation in which ψ holds. A proposition (3), called a dynamic causal law, saying that performing a in a situation in which φ holds causes l to hold in the successor situation. A proposition (4), called a static causal law, states that l holds in any situation in which ϕ holds. A knowledge proposition (5) states that the values of literals in θ, sometimes referred to as sensed-literals, will be known after a is executed.

A *planning problem instance* is a triple π= (D, I, C) where D is a set of propositions of types from (2) to (5), called domain description; I is a set of propositions of type (1), dubbed initial situation; and C is a conjunction of fluent literals.

With the presence of sensing actions we need to extend the notion of plans from a sequence of actions so as to allow conditional statements of the form case-endcase (which subsumes the if-then statement). A conditional plan can be empty, i.e. containing no action, denoted by []; or sequence [a; p] where a is a non-sensing action and p is a conditional plan; or conditional sequence $[a;\ cases(\{gj\ \rightarrow\ pj\}_{j=1}^{n}]$ where a is a sensing action of a proposition (5) with $\theta\ =\ \{g1, \dots, gn\}$ and j 's are conditional plans; Nothing else is a conditional plan.

To execute a conditional plan of the form $[a;\ cases(\{gj\ \rightarrow\ pj\}_{j=1}^{n}]$, we first execute a and then evaluate each gj w.r.t. our current knowledge. If one of the gj's, say gk holds, we execute pk.

ASCP planner works by transforming a given planning problem instance into an ASP program whose answer sets correspond to conditional plans of the problem instance (see [Tu et al. 2007] for details).

2.7.1 Representation in the action language

We now show how the Crow represents Fox's actions language and two problem instances corresponding to the two Fox's intentions, gathered from the CBN: Food (not to be hungry) and Territory (occupy Crow's tree) in Ac K. The representation is inspired by the work in [Kowalski].

Example 1.4 (Fox-Crow (cont'd)). The scenarios with intentions of getting food and territory are represented in Fig1.5 and Fig1.6, respectively. The first problem instance has the conditional plan:

```
[praise(f ox, crow), cases({
        accepted(crow) → [sing(crow), grab(f ox, cheese), eat(fox, cheese)];
        declined(crow) → ⊥})]
```

where ⊥ means no plans appropriate

```
1.  animal(fox). bird(crow). object(cheese). edible(cheese).
    animal(X) :- bird(X).
2.  executable(eat(A,E), [holds(A,E)])        :- animal(A) ,edible(E).
    executable (sing(B), [accepted(B)])       :- bird(B).
    executable(praise(fox,A), []) :- animal(A).
    executable(grab(A,O), [holds(nobody,O)]) :- animal(A) ,object(O).
3.  causes(sing(B),holds(nobody,O), [holds(B,O)])      :- bird(B),object(O).
    causes(eat(A,E) ,neg(hungry(A)), [hungry(A)])      :- animal(A) edible(E).
    causes(grab(A,O),holds(A,O), [])                   :- animal(A),object(O)
4.  determines(praise(fox,B), [accepted(B),declined(B)]) :- bird(B).
5.  initially(holds(crow,cheese)).                     initially(hungry(fox)).
6.  goal([neg(hungry(f ox))]).
```

Fig. 1.5 Fox's plans for food

i.e. first, Fox praises Crow. If Crow accepts to sing, Fox grabs the dropped cheese and eats it. Otherwise, i.e. Crow declines to sing, nothing happens. The second problem instance has the conditional plan:

```
[praise(fox, crow), cases({
        accepted(crow) → [sing(crow), approach(fox, crow), attack(f ox, crow)];
        declined(crow) → ⊥})]
```

Thus, with the only current observation (Fox praised) Crow cannot decide which is the real intention of Fox. Since the only way to identify is an acceptance to sing, which in both cases leads to a bad consequence, losing the cheese and/or the territory, Crow can simply decline to sing. However, being really smart and extremely curious, she can first eat or hide the cheese in order to prevent it from falling down when singing, then she starts singing, keeping an eye on Fox's behaviors. If Fox approaches her, she flies, knowing Fox's real intention is to get her territory (supposing Crow does not get injured by a Fox attack, she can

```
1. place (tree).
2. executable(attack(fox,A), []) :- bird(A), near(fox,A).
   executable (approach(fox,A), [happy(A)]) :- animal(A).
3. causes(attack(fox,A) ,occupied(fox,P), [occupied(A,P)]) :-
                 animal(A),place(P).
   causes(approach(A,B),near(A,B), []) :- animal(A),animal(B).
   causes(sing(A),happy(A), []) :- bird(A).
4. occupied(crow, tree).
5. goal([occupied(fox,tree)]).
```

Fig. 1.6 Fox's plan for territory

revenge on Fox to get back the territory). Otherwise, if Fox does nothing or simply goes away, Crow knows that Fox's real intention was to get the cheese.

3 Evolution Prospection

The next step, that of taking advantage of the recognized intentions gleaned from the previous stage, is implemented using Evolution Prospection Agent (EPA) system [Pereira and Anh 2009b, Pereira and Anh 2009c]. It enables an agent to look ahead, prospectively, into its hypothetical futures, in order to determine the best courses of evolution that satisfy its goals, and thence to prefer amongst those futures. The intentions are provided to the EPA system as goals, and EPA can help with generating the courses of evolution that achieve the goals. These courses of evolution can be provided to the intending agent as suggestions to achieve its intention (in cooperating settings) or else as a guide to prevent that agent from achieving it (in hostile settings).

We next briefly present the constructs of EPA that are necessary for the examples in this article. The whole discussion can be found in [Pereira and Anh 2009b, Pereira and Anh 2009c, Anh et al. 2009].

3.1 Preliminary

We next describe constructs of the evolution prospection system that are necessary for representation of the example. A full presentation can be found in [Pereira and Anh 2009b]. The separate formalism for expressing actions can be found in [Pereira and Anh 2009a] or [Tu et al. 2007].

3.1.1 Language

Let L be a first order language. A domain literal in L is a domain atom A or its default negation $not\ A$. The latter is used to express that the atom is false by default (Closed World Assumption). A domain rule in L is a rule of the form:

$$A \leftarrow L_1, \ldots, L_t (t \geq 0)$$

where A is a domain atom and $L_1, \ldots, L_t$ are domain literals. An integrity constraint in L is a rule with an empty head. A (logic) program P over L is a set of domain rules and integrity constraints, standing for all their ground instances.

3.1.2 Active Goals

In each cycle of its evolution the agent has a set of active goals or desires. We introduce the on observe/1 predicate, which we consider as representing active goals or desires that, once triggered by the observations figuring in its rule bodies, cause the agent to attempt their satisfaction by launching all the queries standing for them, or using preferences to select them. The rule for an active goal AG is of the form:

$$on\ observe(AG) \leftarrow L_1, \ldots, L_t\ (t \geq 0)$$

where $L1, \ldots, L_t$ are domain literals. During evolution, an active goal may be triggered by some events, previous commitments or some history-related information. When starting a cycle, the agent collects its active goals by finding all the $on\_observe(AG)$ that hold under the initial theory without performing any abduction, then finds abductive solutions for their conjunction.

3.1.3 Preferring abducibles

Every program P is associated with a set of abducibles $A \subseteq t$. These, and their default negations, can be seen as hypotheses that provide hypothetical solutions or possible explanations to given queries. Abducibles can figure only in the body of program rules. An abducible A can be assumed only if it is a considered one, i.e. if it is expected in the given situation, and, moreover, there is no expectation to the contrary

$$consider(A) \leftarrow expect(A), not\ expect\_not(A), A$$

The rules about expectations are domain-specific knowledge contained in the theory of the program, and effectively constrain the hypotheses available in a situation. To express preference criteria among abducibles, we envisage an extended language L^*. A preference atom in L^* is of the form a < b, where a and b are abducibles. It means that if b can be assumed (i.e. considered), then a / b forces a to be assumed too if it can. A preference rule in L^* is of the form:

$$a \lhd b \leftarrow L_1,\ldots,L\ (t \geq 0)$$

where $L_1,\ldots,L_t$ are domain literals over L^*. This preference rule can be coded as follows:

$$expect\ not(b) \leftarrow L_1,\ldots,L_n, not\_expect\ not(a), expect(a)\ , not\ a$$

In fact, if b is considered, the consider–rule for abducible b requires $expect\_not(b)$ to be false, i.e. every rule with the head $expect\_not(b)$ cannot have a true body. Thus, $a \lhd b$, that is if its body in the preference rule holds, and if a is expected, and not counter-expected, then a must be abduced so that this particular rule for $expect\_not(b)$ also fails, and the abduction of b may go through if all the other rules for $expect\_not(b)$ fail as well.

A *priori* preferences are used to produce the most interesting or relevant conjectures about possible future states. They are taken into account when generating possible scenarios (abductive solutions), which will subsequently be preferred amongst each other a posteriori.

3.1.4 A posteriori Preferences

Having computed possible scenarios, represented by abductive solutions, more favorable scenarios can be preferred a posteriori. Typically, *a posteriori* preferences are performed by evaluating consequences of abducibles in abductive solutions. An *a posteriori* preference has the form:

$$A_i \ll A_j \leftarrow holds\_given(L_i, A_i), holds\ given(L_j, A_j)$$

where A_i, A_j are abductive solutions and L_i, L_j are domain literals. This means that A_i is preferred to A_j a posteriori if L_i and L_j are true as the side-effects of abductive solutions A_i and A_j, respectively, without any further abduction when testing for the side-effects. Optionally, in the body of the preference rule there can be any Prolog predicate used to quantitatively compare the consequences of the two abductive solutions.

3.1.5 Evolution result a posteriori preference

While looking ahead a number of steps into the future, the agent is confronted with the problem of having several different possible courses of evolution. It needs to be able to prefer amongst them to determine the best courses from its present state (and any state in general). The *a posteriori* preferences are no longer appropriate, since they can be used to evaluate only one-step-far consequences of a commitment. The agent should be able to also declaratively specify preference amongst evolutions through quantitatively or qualitatively evaluating the consequences or side-effects of each evolution choice.

A postriori preference is generalized to prefer between two evolutions. An *evolution result a posteriori* preference is performed by evaluating consequences of following some evolutions. The agent must use the imagination (look-ahead capability) and present knowledge to evaluate the consequences of evolving according to a particular course of evolution. An *evolution result a posteriori preference* rule has the form:

$$E_i \lll E_j \leftarrow holds\ in\ evol(L_i, E_i), holds\_in\_evol(L_j, E_j)$$

where Ei, Ej are possible evolutions and Li, Lj are domain literals. This preference implies that Ei is preferred to Ej if Li and Lj are true as evolution history side-effects when evolving according to Ei or Ej, respectively, without making further abductions when just checking for the side-effects. Optionally, in the body of the preference rule there can be recourse to any Prolog predicate, used to quantitatively compare the consequences of the two evolutions for decision making.

Example 1.5 (Fox-Crow, cont'd). Suppose in Fox-Crow Example 1.4, the final confirmed Fox's intention is that of getting food. Having recognized Fox's hidden intention, what will Crow do to prevent Fox from achieving it? The following EPA program in Fig. 1.7 helps Crow with that.

There are two possible ways so as not to lose the Food to Fox, either simply decline to sing (but thereby missing the pleasure of singing) or hide or eat the cheese before singing.

Line 1 is the declaration of program abducibles (the last two abducibles are for use in the second phase). All of them are always expected (line 2). The counter-expectation rule in line 4 states that an animal is not expected to eat if he is full. The integrity constraints in line 5 say that Crow cannot decline to sing and sing, hide and eat the cheese, at the same time. The *a priori* preference in line 6 states that eating the cheese is always preferred to hiding it (since it may be stolen), of course, just in case eating is a possible solution (this is assured in our semantics of *a priori* preference (sub-section 3.1.3)).

```
1.  abds([decline/0,sing/0,hide/2, eat/2, has_food/0,find_new_food/0]).
2.  expect(decline). expect(sing). expect(hide(_,_)). expect(eat(_,_)).
3.  on_observe(not_losing_cheese) <- has_intention(f ox, food).
    not_losing_cheese <- decline.
    not_losing_cheese <- hide(crow,cheese), sing.
```

```
     not_losing_cheese <- eat(crow,cheese), sing.
4.   expect_not(eat(A,cheese)) <- animal(A), full(A).
     animal(crow).
5.   <- decline, sing.      <- hide(crow,cheese), eat(crow,cheese).
6.   eat(crow,cheese) <| hide(crow,cheese).
7.   no_pleasure <- decline.        has_pleasure <- sing.
8.   Ai << Aj <- holds_given(has_pleasure,Ai), olds_given(no_pleasure,Aj).

9.   on_observe(feed_children) <- hungry(children).
     feed_children <- has_food.  feed_children <- find_new_food.
     <- has_food, find_new_food.
10.  expect(has_food) <- decline, not eat(crow,cheese).
     expect(has_food) <- hide(crow,cheese), not stolen(cheese).
     expect (find_new_food).
11.  Ei <<< Ej <-          hungry(children), holds_in_evol(had_food,Ei),
             holds_in_evol(find_new_food,Ej).
12.  Ei <<< Ej <-holds_in_evol(has_pleasure,Ei),
             holds_in_evol (no_pleasure, Ej).

     beginProlog.
             :- assert(scheduled_events(1, [has_intention(fox,food)])),
             assert(scheduled_events(2, [hungry(children)])).
     endProlog.
```

Fig. 1.7 In Case Fox has intention Food

Suppose Crow is not full. Then, the counter expectation in line 4 does not hold. Thus, there are two possible abductive solutions: *[decline]* and *[eat (crow, cheese), sing]* (since the a priori preference prevents the choice containing *hiding*).

Next, the *a posterori* preference in line 8 is taken into account and rules out the abductive solution containing *decline* since it leads to having *no pleasure* which is less preferred to has *pleasure* – the consequence of the second solution that contains *sing* (line 7). In short, the final solution is that Crow eats the cheese then sings, without losing the cheese to Fox and having the pleasure of singing.

Now, let us consider a smarter Crow who is capable of looking further ahead into the future in order to solve longer term goals. Suppose that Crow knows that her children will be hungry later on, in the next stage of evolution (line 9); eating the cheese right now would make her have to find new food for the hungry children. Finding new food may take long, and is always less favourable than having food ready to feed them right away (evolution result a posteriori preference in line 11). Crow can see three possible evolutions: [[*decline*], [*had food*]]; [[*hide(crow, cheese), sing*], [*had food*]] and [[*eat(crow, cheese), sing*], [*find-new-food*]]. Note that in looking ahead at least two steps into the future, local preferences are not taken into account only after all evolution one were applied (full discussion can be found in [Pereira and Anh 2009b, Anh et al. 2007]).

Now the two *evolution result* a *psoterirori* preferences in lines 11-12 are taken into account. The first one rules out the evolution including *finding new food* since it is less preferred than the other two which includes *had food*. The second one rules out the one including *decline*. In short, Crow will hide the food to keep for her hungry children, and still take pleasure from singing.

Note future events, like *hungry (children)*, can be asserted as Prolog code using the reserved predicate *scheduled events/2*. For more details of its use see [Pereira and Anh 2009b, Pereira and Anh 2009c].

Fig. 1.8 Fox and Crow Fable[1]

4 Intention Recognition and Evolution Prospection for Elder Care

In the last twenty years there has been a significant increase of the average age of the population in most western countries and the number of elderly people has been and will be constantly growing. For this reason there has been a strong development of supportive technologies for elderly people living independently in their own homes, for example, RoboCare Project [Cesta and Pecora 2005] – an

[1] The picture is taken from http://mythfolklore.net/aesopica/bewick/51.htm

ongoing project developing robots for assisting elderly people's living, SINDI – a logic-based home monitoring system [Mileo et Al. 2008], ILSA – an agent-based monitoring and support system for elderly [Haigh 2004] and PHATT – a framework developed for addressing a number of desired features for Elder Care domain [Geib 2002].

For the Elder Care application domain, in order to provide contextually appropriate help for elders, it is required that the assisting system have the ability to observe the actions of the elders, recognize their intentions, and then provide suggestions on how to achieve the recognized intentions on the basis of the conceived plans. The first step of perceiving elders' actions are taken for granted. The second and third steps are addressed by our described Intention Recognition and Evolution Prospection systems, respectively. We include here an example from our previous work [Pereira and Anh 2009d, Anh 2009] for self-containment.

Example 1.6 (Elder Care). An elder stays alone in his apartment. The intention recognition system observes that he is looking for something in the living room. In order to assist him, the system needs to figure out what he intends to find. The possible things are: something to read (*book*); something to drink (*drink*); the TV remote control (*Rem*); and the light switch (*Switch*). The CBN representing this scenario is that of Fig. 1.9.

4.1 Elder Intention Recognition

To begin with, we need to declare two sorts:

```
bool = {t,f}.          elder_intentions = {book,drink,rem,switch}.
```

where the second one is the sort of possible intentions of the elder. There are five top nodes, named *thirsty (thsty)*, *like_reading(lr)*, *like_watching (lw)*, *tv_on(tv)*, *light_on(light)*, belonging to the pre-intention level to describe the causes that might give rise to the considered intentions. The values of last two nodes are observed (evidence nodes). The corresponding random attributes are declared as

```
thsty:bool.          lr:bool.  lw:bool.  tv:bool.  light:bool.
random(rth,thsty,full).        random(rlr, lr,full).          random(rlw,lw,full).
random(rtv,tv,full).           random(rl, light, full).
```

and their independent probability distributions are encoded with pa-rules as

```
thsty:bool.          lr:bool.  lw:bool.  tv:bool.  light:bool.
random(rth,thsty,full).        random(rlr, lr,full).          random(rlw,lw,full).
```

```
random(rtv,tv,full).          random(rl, light, full).
```

The possible intentions reading is afforded by four nodes, representing the four possible intentions of the elder, as mentioned above. The corresponding random attributes are coded specifying an attribute with domain elder intentions and receives boolean values

```
i:elder_intentions --> bool. random(ri, i(I), full).
```

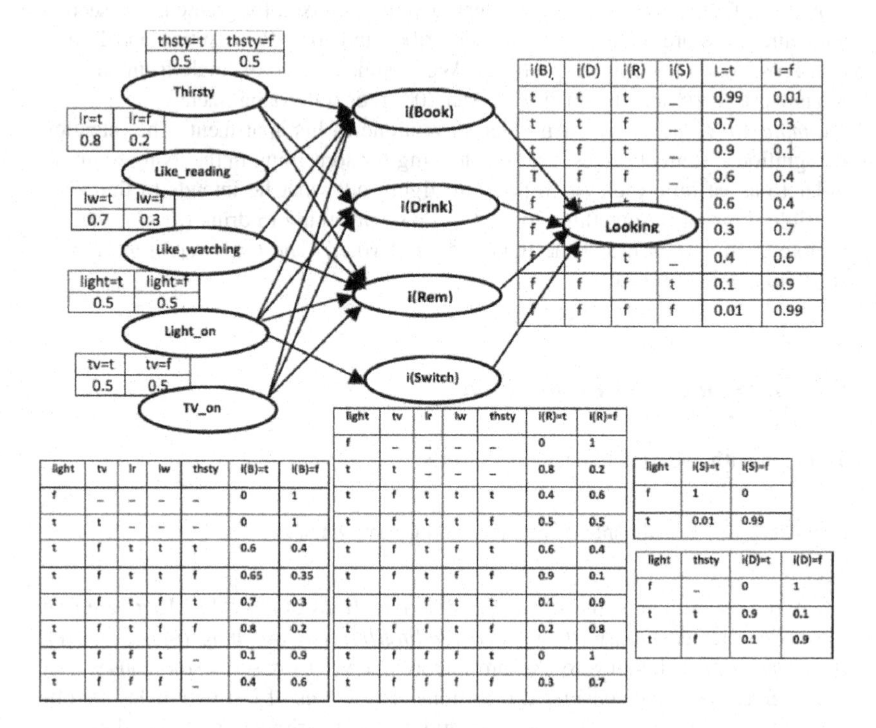

thsty=t	thsty=f
0.5	0.5

lr=t	lr=f
0.8	0.2

lw=t	lw=f
0.7	0.3

light=t	light=f
0.5	0.5

tv=t	tv=f
0.5	0.5

i(B)	i(D)	i(R)	i(S)	L=t	L=f
t	t	t	_	0.99	0.01
t	t	f	_	0.7	0.3
t	f	t	_	0.9	0.1
T	f	f	_	0.6	0.4
f	t	[illegible]	[illegible]	0.6	0.4
f	[illegible]	[illegible]	[illegible]	0.3	0.7
f	f	t	_	0.4	0.6
f	f	f	t	0.1	0.9
f	f	f	f	0.01	0.99

light	tv	lr	lw	thsty	i(B)=t	i(B)=f
f	_	_	_	_	0	1
t	t	_	_	_	0	1
t	f	t	t	t	0.6	0.4
t	f	t	t	f	0.65	0.35
t	f	t	f	t	0.7	0.3
t	f	t	f	f	0.8	0.2
t	f	f	t	_	0.1	0.9
t	f	f	f	_	0.4	0.6

light	tv	lr	lw	thsty	i(R)=t	i(R)=f
f	_	_	_	_	0	1
t	t	_	_	_	0.8	0.2
t	f	t	t	t	0.4	0.6
t	f	t	t	f	0.5	0.5
t	f	t	f	t	0.6	0.4
t	f	t	f	f	0.9	0.1
t	f	f	t	t	0.1	0.9
t	f	f	t	f	0.2	0.8
t	f	f	f	t	0	1
t	f	f	f	f	0.3	0.7

light	i(S)=t	i(S)=f
f	1	0
t	0.01	0.99

light	thsty	i(D)=t	i(D)=f
f	_	0	1
t	t	0.9	0.1
t	f	0.1	0.9

Fig. 1.9 Elder's intentions CBN

The probability distributions of each intention node conditional on the causes are coded in P-log below. Firstly, for $i(book)$:

```
pa(ri(book),i(book,t),d_(0,1))          :-light(f).
pa(ri(book),i(book,t),d_(0,1))          :-light(t),tv(t).
pa(ri(book),i(book,t),d_(6,10))         :-light(t),tv(f),lr(t),lw(t),thsty(t).
pa(ri(book),i(book,t),d_(65,100))       :-light(t),tv(f),lr(t),lw(t),thsty(f).
pa(ri(book),i(book,t),d_(7,10))         :-light(t),tv(f),lr(t),lw(f),thsty(t).
```

```
pa(ri(book),i(book,t),d_(8,10))      :-light(t),tv(f),lr(t),lw(f),thsty(f).
pa(ri(book),i(book,t),d_(1,10))      :-light(t),tv(f),lr(f),lw(t).
pa(ri(book),i(book,t),d_(4,10))      :-light(t),tv(f),lr(f),lw(f).
```

For *i(drink)*:

```
pa(ri(drink),i(drink,t),d_(0,1))     :-light(f).
pa(ri(drink),i(drink,t),d_(9,10))    :-light(t),thsty(t).
pa(ri(drink),i(drink,t),d_(1,10))    :-light(t),thsty(f).
```

For *i(rem)*:

```
pa(ri(rem),i(rem,t),d_(0,1))  :-light(f).
pa(ri(rem),i(rem,t),d_(8,10)) :-light(t),tv(t).
pa(ri(rem),i(rem,t),d_(4,10)) :-light(t),tv(f),lw(t),lr(t),thsty(t).
pa(ri(rem),i(rem,t),d_(5,10)) :-light(t),tv(f),lw(t),lr(t),thsty(f).
pa(ri(rem),i(rem,t),d_(6,10)) :-light(t),tv(f),lw(t),lr(f),thsty(t).
pa(ri(rem),i(rem,t),d_(9,10)) :-light(t),tv(f),lw(t),lr(f),thsty(f).
pa(ri(rem),i(rem,t),d_(1,10)) :-light(t),tv(f),lw(f),lr(t),thsty(t).
pa(ri(rem),i(rem,t),d_(2,10)) :-light(t),tv(f),lw(f),lr(t),thsty(f).
pa(ri(rem),i(rem,t),d_(0,1))  :-light(t),tv(f),lw(f),lr(f),thsty(t).
pa(ri(rem),i(rem,t),d_(3,10)) :-light(t),tv(f),lw(f),lr(f),thsty(f).
```

For *i(switch)*:

```
pa(ri(switch),i(switch,t),d_(1,1))     :-light(f).
pa(ri(switch),i(switch,t),d_(1,100))   :-light(t).
```

There is only one observation, namely, that is the elder is looking for something (look). The declaration of the corresponding random attribute and its probability distribution conditional on the possible intentions are given as follows:

```
look : bool.          random(rla, look, full).
pa(rla,look(t),d_(99,100))     :-i(book,t),i(drink,t),i(rem,t).
pa(rla,look(t),d_(7,10))       :-i(book,t) i(drink,t),i(rem,f).
pa(rla,look(t),d_(9,10))       :-i(book,t),i(drink,f),i(rem,t).
pa(rla,look(t),d_(6,10))       :-i(book,t),i(drink,f),i(rem,f).
pa(rla,look(t),d_(6,10))       :-i(book,f),i(drink,t),i(rem,t).
pa(rla,look(t),d_(3,10))       :-i(book,f),i(drink,t), i(rem,f).
pa(rla,look(t),d_(4,10))       :-i(book,f),i(drink,f),i(rem,t).
pa(rla,look(t),d_(1,10))       :-i(book,f),i(drink,f),i(rem,f),i(switch,t).
pa(rla,look(t),d_(1,100))      :-i(book,f),i(drink,f),i(rem,f),i(switch,f).
```

Recall that the two nodes *tv_on* and *light_ on* are observed. The probabilities that the elder has the intention of looking for *book*, *drink*, *remote control* and *light switch* given the observations that he is looking around and of the states of the light (on or off) and TV (on or off) can be found in P-log with the following queries, respectively:

```
?- pr(i(book, t) | (obs(tv(S1))&obs(light(S2))&obs(look(t))),V1).
?- pr(i(drink, t) | (obs(tv(S1))&obs(light(S2))&obs(look(t))),V2).
?- pr(i(rem, t) | (obs(tv(S1))&obs(light(S2))&obs(look(t))),V3).
?- pr(i(switch, t) | (obs(tv(S1))&obs(light(S2))&obs(look(t))),V4).
```

where S_1, S_2 are boolean values (t or f) instantiated during execution, depending on the states of the light and TV. Let us consider the possible cases

– If the light is off (*S2* = *f*), then *V1* = *V2* = *V3* = 0, *V4* = 1.0, regardless of the state of the TV.
– If the light is on and TV is off (*S1* = *t*, *S2* = *f*), then *V1* = 0.7521, *V2* = 0.5465, *V3* = 0.5036, *V4* = 0.0101.
– If both light and TV are on (*S1* = t, *S2* = t), then *V1* = 0, *V2* = 0.6263, *V3* = 0.9279, *V4* = 0.0102.

Thus, if one observes that the light is off, definitely the elder is looking for the light switch, given that he is looking around. Otherwise, if one observes the light is on, in both cases where the TV is either on or off, the first three intentions book, drink, remote control still need to be put under consideration in the next phase, generating possible plans for each of them. The intention of looking for the light switch is very unlikely to be the case comparing with other three, thus being ruled out. When there is light one goes directly to the light switch if the intention is to turn it off, without having to look for it.

Situation-sensitive CBNs. In this scenario, the CBN may vary depending on some observed factors, for example, the time of day, the current temperature, etc. We design a logical component for the CBN to deal with those factors:

```
pa_rule(pa(rlk,lr(t),d_(0,1)),[])          :-time(T), T>0, T<5, !.
pa_rule(pa(rlk,lr(t),d_(1,10)), [])        :-time(T), T>=5, T<8, !.
pa_rule(pa(rlw,lw(t) ,d_(9,10)), [])       :-time(T),schedule(T,football),!.
pa_rule(pa(rlw,lw(t),d_(1,10)), [])        :-time(T), (T>23; T<5), !.
pa_rule(pa(rth,thsty(t),d_(7,10)),[])      :-temp(T), T>30, !.
pa_rule(pa(rlk,lr(t),d_(1,10)),[])         :-temp(TM), TM >30, !.
pa_rule(pa(rlw,lw(t),d_(3,10)), [])        :-temp(TM), TM>30, !.
```

When the time and temperature are defined (the assisting system should be aware of such information), they are asserted using predicates $time/1$ and $temp/1$. Note that in this modelling, to guarantee the consistency of the P-log program (there must not be two pa-rules for the same attribute instance with non-exclusive bodies) we consider time with a higher priority than temperature, enacted by using XSB Prolog cut operator, as can be seen in the rlk and rlw cases.

4.2 Evolution Prospection for Providing Suggestions

Having recognized the intention of another agent, EPA system can be used to provide the best courses of evolution for that agent to achieve its own intention. These courses of evolution might be provided to the other agent as suggestions.

In Elder Care domain, assisting systems should be able to provide contextually appropriate suggestions for the elders based on their recognized intentions. The assisting system is supposed to be better aware of the environment, the elders' physical states, mental states as well as their scheduled events, so that it can provide good and safe suggestions, or simply warnings. We continue with the Elder Care example from a previous section for illustration.

Example 1.7 (Elder Care, cont 'd). Suppose in Example 1.6, the final confirmed intention is that of looking for a drink. The possibilities are natural pure water, tea, coffee and juice. EPA now is used to help the elder in choosing an appropriate one. The scenario is coded with the program in Fig. 1.10 below.

The elder's physical states are employed in *a priori* preferences and expectation rules to guarantee that only choices that are contextually safe for the elder are generated. Only after that other aspects, for example the elder's pleasure w.r.t. to each kind of drink, are taken into account, in *a posteriori* preferences.

The information regarding the environment (current time, current temperature) and the physical states of the elder is coded in the Prolog part of the program (lines 9-11). The assisting system is supposed to be aware of this information in order to provide good suggestions.

Line 1 is the declaration of program abducibles: $water$, $coffee$, tea, and $juice$. All of them are always expected (line 2). Line 3 picks up a recognized intention verified by the planner. The counter-expectation rules in line 4 state that *coffee* is not expected if the elder has high blood pressure, experiences difficulty to sleep or it is late; and juice is not expected if it is late. Note that the reserved predicate *prolog/1* is used to allow embedding prolog code in an EPA program. More details can be found in [Pereira and Anh 2009b, Pereira and Anh 2009c]. The integrity constraints in line 5 say that is is not allowed to have at the same time the following pairs of drink: tea and coffee, tea and juice, coffee and juice, and tea and water. However, it is the case that the elder can have coffee or juice together with water at the same time.

The *a priori* preferences in line 6 say in the morning coffee is preferred to tea, water and juice. And if it is hot, juice is preferred to all other kinds of drink and water is preferred to tea and coffee (line 7). In addition, the *a priori* preferences in line 8 state if the weather is cold, tea is the most favorable, i.e. preferred to all other kinds of drink.

Now let us look at the suggestions provided by the Elder Care assisting system modelled by this EPA program, considering some cases:

1 time(24) (*late*); temperature(16) (*not hot, not cold*); no high blood pressure; no sleep difficulty: there are two a priori abductive solutions: $[tea]$, $[water]$. Final solutions: $[tea]$ (since it has greater level of pleasure than water, which is ruled out by the a posteriori preference in line 12).

2 time(8) (*morning time*); temperature(16) (*not hot, not cold*); no high blood pressure; no sleep difficulty: there are two abductive solutions: $[coffee]$, $[coffee, water]$. Final: $[coffee]$, $[coffee, water]$.

3 time(18) (*not late, not morning time*); temperature(16) (*not cold, not hot*); no high blood pressure; no sleep difficulty: there are six abductive solutions: $[coffee]$, $[coffee, water]$, $[juice]$, $[juice, water]$, $[tea]$, and $[water]$. Final: $[coffee]$, $[coffee, water]$.

```
1.  abds([water/0, coffee/0, tea/0, juice/0]).
2.  expect(coffee). expect(tea). expect(water). expect(juice).
3.  on_observe(drink)   <- has_intention(elder,drink).
    drink <- tea. drink <- coffee.          drink <- water. drink <- juice.
4.  expect_not(coffee)  <- prolog(blood_high_pressure).
    expect_not(coffee)  <- prolog(sleep_difficulty).
    expect_not(coffee)  <- prolog(late).
    expect_not(juice)   <- prolog(late).
5.  <- tea, coffee.     <- coffee, juice.
    <- tea, juice.      <- tea, water.
6.  coffee <| tea       <- prolog(morning_time).
    coffee <| water     <- prolog(morning_time).
    coffee <| juice     <- prolog(morning_time).
7.  juice <| coffee     <- prolog(hot).     juice <| tea        <- prolog(hot).
    juice <| water      <- prolog(hot).     water <| coffee <- prolog(hot).
    water <| tea        <- prolog(hot).
8.  tea <| coffee       <- prolog(cold).    tea <| juice <- prolog(cold).
    tea <| water        <- prolog(cold).
9.  pleasure_level(3)   <- coffee.          pleasure_level(2) <- tea.
    pleasure_level(1)   <- juice.           pleasure_level(0) <- water.
10. sugar_level(1)      <- coffee.          sugar_level(1) <- tea.
    sugar_level(5)      <- juice.           sugar_level(0) <- water.
11. caffein_level(5)    <- coffee.          caffein_level(0) <- tea.
    caffein_level(0)    <- juice.           caffein_level(0) <- water.
```

```
12. Ai << Aj    <-          holds_given(pleasure_level(V1), Ai),
           holds_given(pleasure_level(V2), Aj), V1 > V2.

13. on_observe (health_check) <- time_for_health_check.
    health_check <- precise_result.
    health_check <- imprecise_result.
14. expect(precise_result) <- no_hight_sugar, no_high_caffein.
    expect (imprecise_result).
    no_high_sugar <- sugar_level(L), prolog(L < 2).
    no_high_caffein <- caffein_level(L), prolog(L < 2).
15. Ei <<< Ej <-            holds_in_evol(precise_result, Ei),
           holds_in_evol(imprecise_result, Ej).

    beginProlog.
    :- assert(scheduled_events(1, [has_intention(elder,drink)])),
           assert(scheduled_events(2, [time_for_health_check])).
           late                    :- time(T), (T > 23; T < 5).
           morning_time            :- time(T), T > 7, T < 10.
           hot                     :- temperature(TM), TM > 32.
           cold                    :- temperature(TM), TM < 10.
           blood_high_pressure     :- phys_cal_state(blood_high_pressure).
           sleep_difficulty        :-physical_state(sleep_difficulty).
    endProlog.
```

Fig. 1.10 Elder Care: Suggestion for a Drink

4 time(18) (*not late, not morning time*); temperature(16) (*not cold, not hot*); high blood pressure; no sleep difficulty: there are four abductive solutions: [*juice*], [*juice, water*], [*tea*], and [*water*]. Final: [*tea*].

5 time(18) (*not late, not morning time*); temperature(16) (*not cold, not hot*); no high blood pressure; sleep difficulty: there are four abductive solutions: [*juice*], [*juice, water*], [*tea*], and [*water*]. Final: [*tea*].

6 time(18) (*not late, not morning time*); temperature(8) (*cold*); no high blood pressure; no sleep difficulty: there is only one abductive solution: [*tea*].

7 time(18) (*not late, not morning time*); temperature(35) (*hot*); no high blood pressure; no sleep difficulty: there are two abductive solutions: [*juice*], [*juice,water*]. Final: [*juice*], [*juice,water*].

If the *evolution result a posteriori preference* in line 15 is taken into account and the elder is scheduled to go to the hospital for health check in the second day: the first and the second cases do not change. In the third case: the suggestions are [*tea*] and [*water*] since the ones that have *coffee* or juice would cause high caffein and sugar levels, respectively, which can make the checking result (health) imprecise (lines 13-15). Similarly for all the other cases . . .

Note future events can be asserted as Prolog code using the reserved predicate *schedule events*/2. For more details of its use see [Pereira and Anh 2009b, Pereira and Anh 2009c].

As one can gather, the suggestions provided by this assisting system are quite contextually appropriate. We might elaborate current factors (time, temperature, physical states) and even consider more factors to provide more appropriate suggestions if the situation gets more complicated.

5 Conclusions and Future Work

We have shown a coherent LP-based system addressing the overall process from recognizing intentions of an agent to taking advantage of those intentions in dealing with the agent, either in a cooperating or hostile settings. The intention recognition part is achieved by means of an articulate combination of situation- sensitive CBNs and a plan generator. Based on the situation at hand and a starting CBN default for the problem domain, its situation-sensitive version is dynamically reconfigured, using LP techniques, in order to compute the likelihood of intentions w.r.t. the situation given, then filter out those much less likely than others. The computed likelihoods enable the recognizing agent to focus on the more likely ones, which is especially important for when having to make a quick decision. Henceforth, the plan generator just needs to work on the remaining relevant intentions. In addition, we have shown how generated plans can guide the recognition process: which actions (or their effects) should be checked for whether they were (hiddenly) executed by the intending agent. We have illustrated all these features with the Fox and Crow example.

We have also shown a LP-based system for assisting elderly people based on the described intention recognizer and Evolution Prospection system. The recognizer is to figure out intentions of the elders based on their observed actions or the effects their actions have in the environment. The implemented Evolution Prospection system, being aware of the external environment, elders' preferences and their note future events, is then employed to provide contextually appropriate suggestions that achieve the recognized intention. The system built-in expectation rules and a priori preferences take into account the physical state (health reports) information of the elder to guarantee that only contextually safe healthy choices are generated; then, information such as the elders pleasure, interests, scheduled events, etc. are taken into account by *a posteriori* and *evolution result a posteriori* preferences.

We believe to have exhibited the usefulness and advantage of our approach of combining several needed features to tackle the Elder Care application domain, by virtue of an integrated logic programming approach. One future direction is to implement meta-explanation about evolution prospection [Pereira et Al. 2009]. It would be quite useful in the considered setting, as the elder care assisting system should be able to explain to elders the whys and wherefores of suggestions made. Moreover, it should be able to produce the abductive solutions found for possible evolutions, keeping them labeled by the preferences used (in a partial order) instead of exhibiting only the most favorable ones. This would allow for final preference change on the part of the elder.

There are currently several other possible future directions to explore. First of all, we can employ an interplay between CBNs and the planner. Besides being a consumer of CBNs as shown, the planner can also be a producer for the CBN in the following ways. Firstly, its feedback about the plausible final intention of the intending agent may increase the corresponding probabilistic relations of the confirmed intention in the CBN; secondly, when new actions (or their effects) of the intending agent, not observed before, become confirmed, the CBN is updated again, which might rule out more intentions, not yet explored nor able to be confirmed or denied. Moreover, the planner might do real experiments, or even thought experiments, where values of nodes may be enforced true. The thought experiments may involve hypothetical or even counterfactual reasoning (possibly prospecting the future [Pereira and Anh 2009b]).

In addition, the advance in LP semantics for evolving program with updates [Alferes et Al. 2002] should be used to give more flexibility in updating CBNs with new information. This is essential when more dynamic reasoning processes, e.g. in the above CBNs-Planner interplay, are employed. To this end, we also plan to parameterize P-log, i.e. enable P-log to have variables in different constructs, such as sort declarations, probabilistic information pa-rules, etc., and those variables be provided by the program calling it, depending on the context. CBNs updating would very much benefit from this ability.

Clearly, an agent recognizes the intention of another for a purpose, i.e. Intention Recognition should be purposive. The depth of understanding of an intention that is required by an agent depends on why knowledge of the intention is required. It might be that an agent needs only the broadest understanding of the intention. In that case, simply knowing the general class of intention is adequate and details are unimportant. But it also might be that details are important for a given purpose. For example, in the Elder Care example (Example 1.6), confirming that the elder has an intention of looking for something is not enough; the details about what he/she is looking for are necessary for the purpose of providing appropriate support or suggestions. To this end, we plan to use an ontology of intentions, and the more general intentions are discovered earlier. Confirmation of a general intention may trigger the discovery of more specific ones, if more details of understanding of the intention are required and available. Actually, richer details of understanding an intention might come up during the recognition process of a general intention as more observations can be gathered. An example of this arises in the Fox-Crow example (Example 1.1). Initially, the system is trying to recognize a general intention: if Fox intends to get some food; and during the recognition process, the detail that Fox's intention is to get a concrete kind of food, Crow's cheese, is found out. However, this is not always the case. For instance, in the Elder Care example, confirming that the elder is looking for something simply triggers a new Intention Recognition process, including the design of a new CBN for computing the likelihood of specific intentions (a drink, a book, TV remote control or light switch) and generating plans for the likely ones, so as to figure out more details about the intention. We plan to attempt to categorize the possible cases and conduct appropriate techniques for each of them.

References

Alferes J, Brogi A, Leite J (2005) The Refined Extension Principle for Semantics of Dynamic Logic Programming, Studia Logica 79(1): 7-32

Alferes J, Brogi A, Leite J, Pereira L Evolving logic programs. Procs. 8th Europ. Conf. on Logics in AI (JELIA'02), pp. 50–61, Springer LNAI 2424

Alferes J, Brogi A, Pereira L Przymusinska H, Przymusinski T (2000) Dynamic updates of non-monotonic knowledge bases. J. Logic Programming, 45(1- 3):4370

An implementation of ASCP using XASP available at: http://centria.di.fct.unl.pt/ lmp/software/cataplan-online.zip

Anh H (2009) Evolution Prospection with Intention Recognition via Computational Logic. Master Thesis Technical University of Dresden

Anh H, Ramli C, Damásio C (2008) An implementation of extended P-log using XASP, in: M. Garcia de la Banda, E. Pontelli (eds.), In Procs. Intl. Conf. Logic Programming, pp. 739–743, Springer LNCS 5366

Baral C, Gelfond M, Rushton N (2004). Probabilistic reasoning with answer sets. In Procs. Logic Programming, Nonmonotonic Reasoning (LPNMR 7), pages 21–33, Springer LNAI 2923, 2004.

Baral C, Gelfond M, Rushton N (2009) Probabilistic reasoning with answer sets. Theory, Practice of Logic Programming, 9(1): 57-144

Castro L, Swift T, Warren D XASP: Answer set programming with xsb and smodels. Accessed at http://xsb.sourceforge.net/packages/xasp.pdf

Cesta A, Pecora F (2005) The Robocare Project: Intelligent Systems for Elder Care. AAAI Fall Symposium on Caring Machines: AI in Elder Care, USA

Eiter T, Faber W, Leone N, Pfeifer G, Polleres A (2003) A Logic Programming Approach to Knowledge State Planning, II: The DL V/C System. Artificial Intelligence 144(1-2): 157-211

Geib W (2002) Problems with intent recognition for elder care. In Procs. AAAI Workshop Automation as Caregiver

Gelfond M, Lifschitz V (1993) Representing actions and change by logic programs. Journal of Logic Programming 17, 2,3,4, 301–323

Giuliani M, Scopelliti M, Fornara F (2005) Elderly people at home: technological help in everyday activities. IEEE International Workshop on In Robot and Human Interactive Communication pp. 365-370

Glymour C (2001) The Mind's Arrows: Bayes Nets and Graphical Causal Models in Psychology. MIT Press

Haigh K, Kiff L, Myers J, Guralnik V, Geib C, Phelps J, Wagner T (2004) The independent lifestyle assistant (i.l.s.a.): Ai lessons learned. In Procs. of Conf. on Innovative Applications of AI, 852857

Heinze C (2003) Modeling Intention Recognition for Intelligent Agent Systems Doctoral Dissertation, the University of Melbourne, Australia

Kautz H, Allen J (1986) Generalized plan recognition. In Procs. 1986 Conf. of the American Association for Artificial Intelligence, AAAI 1986: 32-37

Kowalski. How to be Artificially Intelligent, online book. Downloadable at: http://www.doc.ic.ac.uk/ rak/

Mileo A, Merico D, Bisiani R (2008) A Logic Programming Approach to Home Monitoring for Risk Prevention in Assisted Living, ICLP, Springer LNCS 5366

Niemelä, Simons P (1997) Smodels: An implementation of the stable model and well- founded semantics for normal logic programs. 4th Intl. Conf. on Logic Programming and Nonmonotonic Reasoning Springer LNAI 1265 pages 420–429

Pearl J (2000) Causality: Models, Reasoning, Inference. Cambridge U.P.

Pereira L, Anh H (2009a) Intention Recognition via Causal Bayes Networks plus Plan Generation, in: Seabra Lopes L.; Lau N.; Mariano P.; Rocha L. (eds.) Progress in Artificial Intelligence, Procs. 14th Portuguese Intl.Conf. on Artificial Intelligence (EPIA'09), pp. 138-149, Springer LNAI 5816

Pereira L, Anh H (2009b) Evolution Prospection, in: K. Nakamatsu (ed.), Procs. Intl. Symposium on Intelligent Decision Technologies (KES-IDT'09), pages 51-63, Springer Studies in Computational Intelligence 199

Pereira L, Anh H (2009c) Evolution Prospection in Decision Making, in: Intelligent Decision Technologies (IDT), 3(3):157–171

Pereira L, Anh H (2009d) Elder Care via Intention Recognition and Evolution Prospection. in: S. Abreu, D. Seipel (eds.), Procs. 18th Intl. Conf. on Applications of Declarative Programming, Knowledge Management (INAP'09)

Pereira L, Pinto A (2009) Inspection Points and Meta-Abduction in Logic Programs, in: S. Abreu, D. Seipel (eds.), Procs. 18th Intl. Conf. on Applications of Decl. Programming and Knowledge Management (INAP'09), pp. 171–184

Schrempf O Albrecht D Hanebeck U (2007) Tractable Probabilistic Models for Intention Recognition Based on Expert Knowledge, In Procs. 2007 IEEE/RSJ Intl. Conf. on Intelligent Robots, Systems (IROS 2007), pages 1429–1434

Swift T (1999) Tabling for non-monotonic programming. Annals of Mathematics and Artificial Intelligence, 25(3–4):201-240

Tahboub K (2006) Intelligent Human-Machine Interaction Based on Dynamic Bayesian Networks Probabilistic Intention Recognition. J. Intelligent Robotics Systems, vol. 45, no. 1, pages 31-52

The XSB System Version 3.0 Vol. 2: Libraries, Interfaces and Packages. July 2000

Tu P, Son T, Baral C (2007) Reasoning and Planning with Sensing Actions, Incomplete Information, and Static Causal Laws using Answer Set Programming. Theory and Practice of Logic Programming, 7(4): 377-450

Scheduling a Cutting and Treatment Stainless Steel Sheet Line with Self-Management Capabilities

Ana Madureira[1,2], Ivo Pereira[1,2], Nelson Sousa[1,2], Paulo Ávila[2] and João Bastos[2]

[1] GECAD – Knowledge Engineering and Decision Support Group Porto, Portugal, *(email :{amd, iasp, nffs}@isep.ipp.pt)*

[2] Institute of Engineering – Polytechnic of Porto, Porto, Portugal, *(email: {psa, jab}@isep.ipp.pt)*

Abstract With advancement in computer science and information technology, computing systems are becoming increasingly more complex with an increasing number of heterogeneous components. They are thus becoming more difficult to monitor, manage, and maintain. This process has been well known as labor intensive and error prone. In addition, traditional approaches for system management are difficult to keep up with the rapidly changing environments. There is a need for automatic and efficient approaches to monitor and manage complex computing systems. In this paper, we propose an innovative framework for scheduling system management by combining Autonomic Computing (AC) paradigm, Multi-Agent Systems (MAS) and Nature Inspired Optimization Techniques (NIT). Additionally, we consider the resolution of realistic problems. The scheduling of a Cutting and Treatment Stainless Steel Sheet Line will be evaluated. Results show that proposed approach has advantages when compared with other scheduling systems.

1 Introduction

Manufacturing scheduling can be defined as the allocation, over the time, of jobs to machines, within a short temporal horizon and according to a specific criterion, such as cost or tardiness. It is a complex combinatorial problem, more specifically a non-polynomial (NP) problem: the objective is to find the optimal sequence from the $j!^m$ possible schedules, where j is the number of jobs and m the number of machines. Manufacturing scheduling domains are characterized by a great amount of uncertainty that leads to significant system dynamism. Such dynamic scheduling is receiving increased attention amongst both researchers and practitioners.

Scheduling is an important aspect of automation in manufacturing systems. Most of scheduling domains are characterized by a great amount of uncertainty

A. Madureira et al. (eds.), *Computational Intelligence for Engineering Systems: Emergent Applications*, Intelligent Systems, Control and Automation: Science and Engineering 46,
DOI 10.1007/978-94-007-0093-2_2, © Springer Science+Business Media B.V. 2011

that leads to significant system dynamics [Aytg et al. 2005]. Such dynamic scheduling is receiving increased attention amongst both researchers and practitioners [Aytug et al. 2005, Madureira 2003, Madureira et al. 2008, Madureira et al. 2007, Wellner and Dilger 1999]. However, scheduling is still having difficulties in real world situations and, hence, human intervention is required to maintain real-time adaptation and optimization.

Dynamic changes of a problem could arise from new user requirements and the evolution of the external environment. In a more general view, dynamic problem changes can be seen as a set of constraint insertions and cancellations.

For these dynamic optimization problems environments, that are often impossible to avoid in practice, the objective of the optimization algorithm is no longer to simply locate the global optimum solution, but to continuously track the optimum in dynamic environments, or to find a robust solution that operates optimally in the presence of perturbations [Aytug et al. 2005, Madureira et al. 2007]. In spite of all the previous trials, the scheduling problem is still known to be NP-complete, even for static environments. This fact poses serious challenges to conventional algorithms and incites researchers to explore new directions [Madureira 2003, Madureira et al. 2008, Madureira et al. 2007] and Multi-Agent technology has been considered an important approach for developing industrial distributed systems.

This paper addresses the use of Multi-Agent Systems paradigm for supporting dynamic and distributed scheduling in Manufacturing Systems with Autonomic properties, in order to reduce the complexity of managing systems and human interference. Additionally, we consider the resolution of realistic problems. The scheduling of a Cutting and Treatment Stainless Steel Sheet Line will be evaluated.

The remaining sections are organized as follows: in Section 2 Nature Inspired Optimization Techniques are presented. Section 3 describes Multi-Agent Systems paradigm. Section 4 summarizes some aspects and related work on Autonomic Computing. In section 5 the AutoDynAgents system is presented. Section 6 presents the description of the Production Process of the Cutting and Treatment Stainless Steel Sheet Line and the computational study. Finally, the paper presents some conclusions and puts forward some ideas for future work.

2 Nature Inspired Optimization Techniques

Many optimization problems in diverse fields have been solved using different optimization algorithms. Traditional optimization techniques such as linear programming (LP), non-linear programming (NLP), and dynamic programming (DP) have had major roles in solving these problems. However, their drawbacks generate demand for other types of algorithms, such as heuristic optimization approaches (Simulated Annealing, Tabu Search, and Evolutionary Algorithms).

However, there are still some possibilities of devising new heuristic algorithms based on analogies with natural or artificial phenomena or even the development of hybrid approaches.

The interest of the NIT based approaches is that they converge, in general, to satisfactory solutions in an effective and efficient way (computing time and implementation effort). NIT have often been shown to be effective for difficult combinatorial optimization problems appearing in various industrial, economical, and scientific domains [Gonzalez 2007, Madureira et al. 2007, Siarry 2008].

When considering and understanding solutions followed by nature it is possible to use this acquired knowledge on the resolution of complex problems on different domains. From this knowledge application, on a creative way, arise new computing science areas – Bionic Computing.

The complexity of current computer systems has led the software engineering, distributed systems and management communities to look for inspiration in diverse fields, e.g. robotics, artificial intelligence or biology, to find new ways of designing and managing systems. Hybridization of different approaches seems to be a promising research field of computational intelligence focusing on the development of the next generation of intelligent systems.

3 Multi-Agent Systems

Multi-agent paradigm is emerging for the development of solutions to very hard distributed computational problems. This paradigm is based either on the activity of "intelligent" agents which perform complex functionalities or on the exploitation of a large number of simple agents that can produce an overall intelligent behavior leading to the solution of alleged almost intractable problems. The multi-agent paradigm is often inspired by biological systems that are based in the Social Systems interactions between agents and subject to negotiations.

Considering the complexity inherent to the manufacturing systems, dynamic scheduling is considered an excellent candidate for the application of agent-based technology.

The main term of Multi-Agent based computing is an Agent. However the definition of the term Agent has not common consent. In the last few years most authors agreed that this definition depends on the domain where agents are used. However there is a general consensus about its two main abstractions:

- An agent is a computational system that is situated in a dynamic environment and is capable of exhibiting autonomous and intelligent behaviour.
- An agent may have an environment that includes other agents. The community of interacting agents, as a whole, operates as a multi-agent system.

Some of most important common properties of computational agents are as follows [Wooldridge 2002]:

- act on behalf of their designer or the user they represent in order to meet a particular purpose.
- are autonomous in the sense that they control both their internal state and behaviour in the environment.
- exhibit some kind of intelligence, from applying fixed rules to reasoning, planning and learning capabilities.
- interact with their environment, and in a community, with other agents.
- are ideally adaptive, i.e., capable of tailoring their behaviour to the changes of the environment without the intervention of their designer.

Additional agent properties, characteristic in particular domains and applications are mobility (when an agent can transport itself to another environment to access remote resources or to meet other agents), genuineness (when it does not falsify its identity), credibility or trustworthiness (when it does not communicate false information wilfully) and sociality (when agents work in open operational environments hosting the execution of a multiplicity of agents, possibly belonging to different stakeholders.

In many implementations of MAS systems for manufacturing scheduling, the agents model the resources of the system and the tasks scheduling is done in a distributed way by means of cooperation and coordination amongst agents [Aytug et al. 2005, Bhat et al. 2006]. There are also approaches that use a single agent for scheduling that defines the schedules that the resource agents will execute [Abdelwahed and Kandasamy 2006, Bustard and Sterritt 2006]. When responding to disturbances, the distributed nature of multi-agent systems can also be a benefit to the rescheduling algorithm by involving only the agents directly affected, without disturbing the rest of the community that can continue with their work.

4 Autonomic Computing

Autonomic Computing is an IBM Grand Challenge proposed in 2001 by Paul Horn, Senior Vice-President of IBM Research [Horn 2001]. Horn argues that the Information Technology (IT) industry focus on constant expansion will soon reach its breaking point: massive data centres are built in organic, ad hoc ways, resulting in a heterogeneous composition whose maintenance costs in terms of qualified staff, time and capital will soon exceed corporate capabilities.

AC proposes a broad new field of research related to the automation of IT management processes, drawing inspiration from the human autonomous nervous system. From its inception, the concept revolves around four self-* properties, in which research efforts may be categorized: Self-Configuring, Self-Healing, Self-Optimizing and Self-Protecting. This number was by no means restricted, but no other proposals seem to have been made thus far.

Although the names of these properties are fairly self-explanatory, there is one inherent and implicit concept of significant importance: proactiveness. This is what separates this area of research from some of the functionalities which are already being integrated with existing software systems.

Software systems managing IT resources without human supervision, called Autonomic Managers, are expected to continuously and autonomously respond to changes, and continuously seek ways to improve efficiency or counter negative environment changes.

Many studies have already been made around this area. These range from Software Engineering concerns to address this new development paradigm [Bustard and Sterritt 2006, Cervenka et al. 2006] all the way down to industry integration [Ganek 2006, IBM 2006].

Important techniques have been tapped into from areas such as Service-Oriented Architectures [Adams et al. 2006, Bhat et al. 2006, Chess et al. 2006] Schwan et al. 2006], Multi-Agent Systems [Kephart and Chess 2003], Grid Computing (Ganek 2006)(Kephart and Chess 2003) and Control Theory [Abdelwahed and Kandasamy 2006, Bhat et al. 2006].

The Organization for the Advancement of Structured Information Standards (OASIS) has also furthered the establishment of standards in the very important area of communication, mostly related to Web Services [Chess et al. 2006]. These have proven to be central in the quest for a communication middleware layer that can effectively abstract away the heterogeneity of underlying IT components.

Planning is a critical component of the Autonomic Computing vision [Kephart and Chess 2001] where in the behavior of system elements are monitored and analyzed, and the performance is used to plan and execute suitable actions to take or keep the system in desirable states. In the AC vision [Kephart and Chess 2001], four aspects of self-* are distinguished:

- *Self-configuration*: deals with installation, configuration and integration of IT systems. The installation procedures work by gathering information about the host environment, figuring out the dependencies among needed tasks and also optimizing performance measures, and finally executing the tasks to realize the changes. Information about host system is increasingly getting standardized along structured formats but the executable tasks can be adhoc scripts. Humans want to be closely involved in key decisions during execution.
- *Self-optimizing*: deals with improving the performance of running systems by leveraging alternative opportunities. The system would monitor its performance and based on its changing environment, could initiate new changes (e.g., resource re-provisioning).
- *Self-healing*: deals with determination of problematic situations and recovering from them. It requires the system to reason with how activities can be performed, how diagnostic information is produced and how new changes can be affected with minimal cost and maximum benefit. The specification of actions could be known at some level of granularity.

- *Self-protecting*: deals with monitoring the environment for threats and responding to them. It is related to self-optimizing aspect but with the difference that the situation needs time-bound response and lead to cascading effect.

5 AutoDynAgents System

AutoDynAgents is an Autonomic Scheduling System in which communities of agents model a real manufacturing system subject to perturbations. Agents must be able to learn and manage their internal behavior and their relationships with other autonomic agents, by cooperative negotiation in accordance with business policies defined by user manager.

The main purpose of AutoDynAgents is a Multi-Agent System where each agent represents a resource (Resource Agents) in a Manufacturing System. Each Resource Agent must be able: to find an optimal or near optimal local solution through Genetic Algorithms, Tabu Search or other NIT; to deal with system dynamism (new jobs arriving, cancelled jobs, changing jobs attributes, etc); to change/adapt the parameters of the basic algorithm according to the current situation; to switch from one Meta-Heuristic algorithm to another and to cooperate with other agents.

Scheduling approach followed by AutoDynAgents system is rather different from the ones found in the literature; as we try to implement a system where each Resource Agent is responsible for optimizing the scheduling of operations for one machine through a NIT. This considers a specific kind of social interaction that is cooperative problem solving (CPS), where the group of agents work together to achieve a good solution for the problem.

The original scheduling problem defined in [Madureira 2003, Madureira et al. 2007], is decomposed into a series of Single Machine Scheduling Problems (SMSP). The Resource Agents (which has an NIT associated) obtain local solutions and later cooperate in order to overcome inter-agent constraints and achieve a global schedule.

Two possible approaches, to deal with this problem, could be used. In the first, the AutoDynAgents system waits for the solutions obtained by the Resource Agents and then apply a repair mechanism to shift some operations in the generated schedules till a feasible solution is obtained (Repair Approach). In the second, a coordination mechanism is established between related agents in the process, in order to interact with each other to pursuit common objective through cooperation. These coordination mechanism must be prepared to accept agents subjected to dynamism (new jobs arriving, cancelled jobs, changing jobs attributes).

The AutoDynAgents system architecture (Fig. 2.1 and 2.2) is based on six different types of agents.

In order to allow a seamless communication with the user, an *User Interface Agent* was implemented. This agent, apart from being responsible for the user interface, generates the necessary Job Agents dynamically according to the number of tasks that comprise the scheduling problem and assign each task to the respective Task Agent.

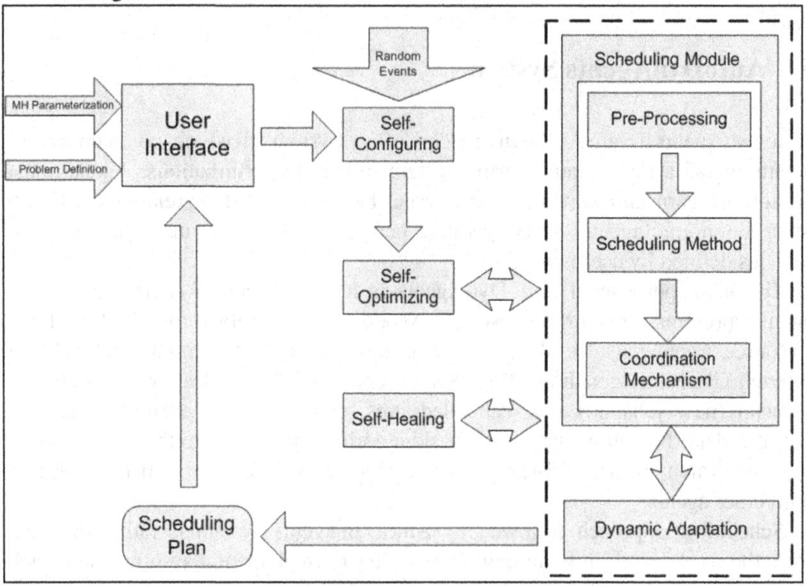

Fig 2.1 AutoDynAgents system architecture

The *Task Agent* will process the necessary information about the job. That is to say that this agent will be responsible for the generation of the earliest and latest processing times, the verification of feasible schedules and identification of constraint conflicts on each job and the decision on which Machine Agent is responsible for solving a specific conflict.

The *Machine Agent* is responsible for the scheduling of the operations that require processing in the machine supervised by the agent. This agent will implement meta-heuristic and local search procedures in order to find best possible operation schedules and will communicate those solutions to the Task Agent for later feasibility check.

Respectively to the Self-*Agents, the *Self-Configuring Agent* is responsible for monitoring the system in order to detect changes occurred in the schedule, allowing the system to a dynamic adaptation. With this agent, the system will be prepared to automatically handle dynamism by adapting the solutions to external perturbations. While, on one hand, partial events only require a redefinition of jobs' attributes and re-evaluation of the objective function, on other hand, total events require changes on the solution's structure and size, carried out by insertion or deletion of operations, and also re-evaluation of the objective function. Therefore, under total events, the modification of the current solution is imperative, through job arrival integration mechanisms (when a new job arrives to be processed), job elimination mechanisms (when a job is cancelled) and regeneration mechanisms in order to ensure a dynamic adaptation of population/neighborhood.

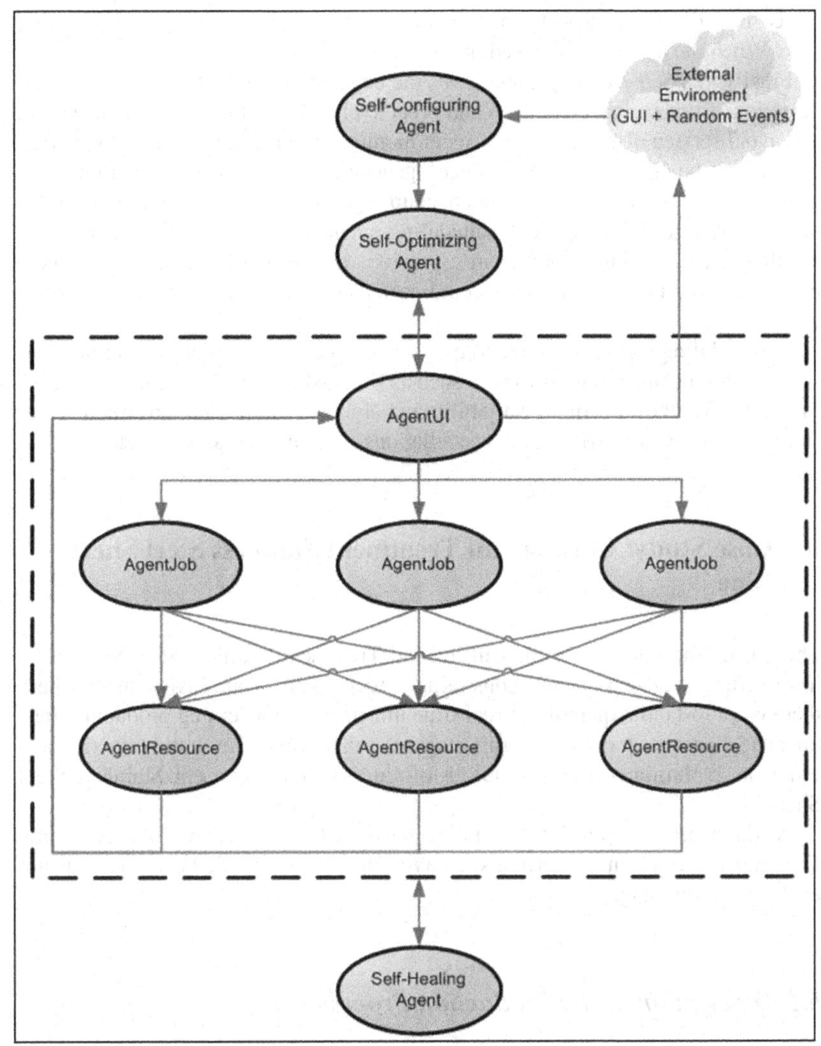

Fig 2.2. AutoDynAgents agents' architecture

The *Self-Optimizing Agent* is responsible for the automatically tuning of the meta-heuristics' parameters, according to the problem. This agent receives the initial problem, or the changes detected by Self-Configuring Agent, and automatically choose the meta-heuristic to use, and makes its self-parameterization. If some dynamism occurs, parameters may change in run-time. This tuning of parameters is made through learning and experience, since it uses a Case-based Reasoning (CBR) module. Each time a new problem (case) appears, the CBR uses past experience in order to specify the meta-heuristic and respective parameters for that case. When the new case is solved, it is stored for later use.

Finally, the *Self-Healing Agent* gives the capacity to the system for diagnosing deviations from normal conditions and proactively takes actions to normalize them and avoid service disruptions. This agent monitors other agents in order to provide overall self-healing capabilities. Since agents may crash for some reason, self-healing provides one or more agents backup registries in order to grant storage for the reactivation of lost or stuck scheduling agents with meaningful results, thus enabling the system to restart from a previous checkpoint as opposed to a complete reset. With this agent, the system becomes stable, even if some deadlocks or crashes occur.

Rescheduling is necessary due to two classes of events: Partial events imply variability in jobs/operations attributes such as processing times, due dates or release times; and Total events imply variability in neighborhood/population structure, resulting from new job arrivals, job cancellations, machines breakdown, etc.

6 Case Study: Cutting and Treatment Stainless Steel Sheet Line

The scheduling systems for the Cutting and Treatment Stainless Steel Sheet Line have different objectives and constraints and operate in an environment where there is a substantial quantity of real-time information concerning production failures and customer requests. At this stage, the main objective is the effective and efficient resolution of the Scheduling of Cutting and Treatment Stainless Steel Sheet Line.

At this work the AutoDynAgents has been evaluated only from its scheduling system perspective on a deterministic environment. Autonomic behaviour was not evaluated at this stage.

6.1 Description of the Production Process

The production process under study is a Job-Shop like scheduling problem, which pretends the processing the cut of plates of stainless steel with or without superficial treatment based on gridding. The number of different final products is very large, i.e., each order is normally different from another, because the enterprise works majority with product specifications (dimensions and type of superficial treatment) by order. By this reason, the enterprise scheduling problem is sequence dependent of setup times and its optimization is very important for the system competitiveness, due to the setup time, which varies from 7 minutes to 18 minutes.

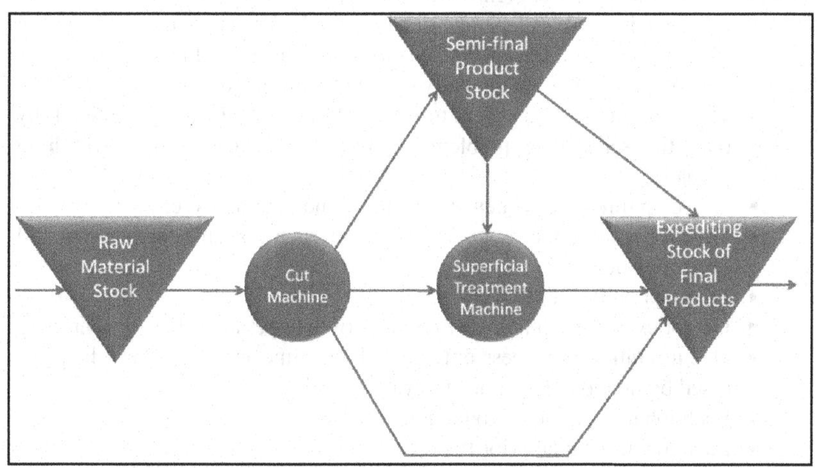

Fig 2.3. Production Process Flow

The production system is constituted by two machines (Fig. 2.3): the cut machine, automatic, and the superficial treatment machine, semi-automatic; and by three storage zones: raw material stock, semi-final product stock, and expediting stock of final products. The routing of the products, showed in the figure below that represents the process chart, can be of four different ways. The principal one is the horizontal, passing through the two machines. The second one begins in the raw material stock, pass through the cut machine and finish in the expediting stock of final products. The third one begins in the raw material stock, pass through the cut machine, then goes through the semi-final product stock and finish in the expediting stock of final products. The fourth one begins in the semi-final product stock, pass through the superficial treatment machine and finish in the expediting stock of final products.

6.2 Scheduling Problem Description

The production orders, more than 10 per day in average, are obtained through the current planning system (ERP) and comprehend the following information: product type, quantity, and production date. After that, the production orders, just allocated to the routing process, are transferred to the scheduling tool, the *Izaro Grey* from the *Softi9* enterprise, that sequence the orders attending the minimization of the *makespan*.

The organization scheduling problem is performed by the *Izaro Grey* based on the following assumptions coherent with its reality:

- All the *n* jobs are independent and available for processing at the initial time, according to the ERP results for a time horizon of nine days;
- Are previously known the delivery date for all the jobs;
- The allocation of the orders to the machines it is already predefined, by that, the scheduling problem resume to the Sequencing/Dispatching problem;
- Job setup times are sequence dependent and deterministic;
- For all the jobs, the processing times at each operation are known and deterministic;
- Pre-emption is not allowed;
- Do not exist precedence restrictions between operations of different jobs;
- One machine can process only one job at a time and one job can be processed by only one machine at any time;
- Each job have a linear production process;
- Each job have equal priority order;
- The production scheduling is static during a period of one to three days;
- The scheduling objective is to minimize the *makespan*;
- The production work station has sufficient capacity to store and manage the work–in-process inventory generated during the execution of the complete set of jobs. That is, we assume infinite capacity at each stage;
- All the three storage zones referred in the routing of the products have sufficient capacity;
- Travel time between consecutive stages is negligible.

According to the literature [Pinedo 2008], the scheduling problem of this enterprise is denoted by $J_m \,||\, C_{max}$, that designates a job shop problem with *m* machines, in our case two. There is no recirculation, the job visits each machine at most once and the objective is the minimization of the *makespan*.

6.3 Simulation Plans and Computational Results

In order to analyse the performance of the AutoDynAgents with the scheduling systems, Izaro APS2, which is used at the enterprise and Lekin3, we selected a pe-

---

[2] http://www.softi9.pt/images/download/doc32.pdf

[3] http://www.stern.nyu.edu/om/software/lekin/index.htm

riod of nine days of production orders (period used by the enterprise to scheduling performing). Considering that Lekin has capacity restrictions in the total number of production orders, we had considered two simulation plans: Plan 1 – Izaro versus AutoDynAgents for all the 90 production orders; Plan 2 – Lekin versus AutoDynAgents for the first 62 production orders.

Table 2.1 Simulation Plans

Simulation Plans	Scheduling System		Number of Production Orders
Plan 1	Izaro	AutoDynAgents	90
Plan 2	Lekin	AutoDynAgents	62

Considering that scheduling systems under analysis are based in stochastic based algorithms, we have run ten simulations for each plan and choose the best solution for each scheduling system. The final results obtained for the each plans are showed in tables 2.2 and 2.3.

Table 2.2 Simulation Results for the Plan 1

Scheduling Tools	Izaro	AutoDynAgents
Makespan (min)	4070	3950
Maximum Lateness (min)	591	0
Number of Late Jobs	8	0
ΣLateness (min)	3018	0
Total Production Time (min)	4165	4305

Table 2.3 Simulation Results for the Plan 2

Scheduling Tools	Lekin	AutoDynAgents
Makespan (min)	2664	2549
Maximum Lateness (min)	0	51
Number of Late Jobs	0	1
ΣLateness (min)	0	51
Total Production Time (min)	2618	2689

When comparing the obtained results by Izaro and AutoDynAgents systems (Table 2.2) we can conclude about the advantage in effectiveness obtained by AutoDynAgents in almost optimization measures, namely on the minimization of *makespan*, maximum *lateness*, number of late jobs and the sum of lateness. In simulation plan 2 (Table 2.3), AutoDynAgents system outperforms Lekin system when analyzed *makespan* optimization criteria, but with a degradation of lateness related optimization criteria. It is possible to conclude about the general advantage in effectiveness of AutoDynAgents over Izaro APS and Lekin scheduling systems.

7 Conclusions

We believe that a new contribution for the resolution of more realistic scheduling problems was described in this paper. A novel approach to scheduling resolution by combining Autonomic Computing, Multi-Agent Systems and Nature Inspired Optimization Techniques was proposed. The use of Multi-Agent Systems paradigm for supporting dynamic and distributed scheduling in Manufacturing Systems with Autonomic properties, in order to reduce the complexity of managing systems and human interference was supported. Additionally, we consider the resolution of realistic problems: the scheduling of a Cutting and Treatment Stainless Steel Sheet Line was evaluated.

The experimental results showed the performance of proposed scheduling on the several plans simulations advantage over the others scheduling systems under considerations.

Work still to be done includes the exhaustive testing of the proposed system and negotiation mechanisms under dynamic environments subject to several random perturbations.

Acknowledgments The authors would like to acknowledge FCT, FEDER, POCTI, POCI for their support to R&D Projects and GECAD - Knowledge Engineering and Decision Support Group Unit.

References

Abdelwahed S, Kandasamy N (2006) A Control-Based Approach to Autonomic Performance Management in Computing Systems, in Autonomic Computing – Concepts, Infrastructure, and Applications

Adams R, Brett P, Iyer S, Milojicic D, Rafaeli S, Talwar V (2006) Scalable Management – Technologies for Management of Large-Scale, Distributed Systems, in Autonomic Computing – Concepts, Infrastructure, and Applications

Aytug H, Lawley MA, McKay K, Mohan S, Uzsoy R (2005) Executing production schedules in the face of uncertainties: A review and some future directions. European Journal of Operational Research, Volume 16 (1), 86-110

Bhat V, Parashar M, Kandasamy N (2006) Autonomic Data Streaming for High-Performance Scientific Applications, in Autonomic Computing – Concepts, Infrastructure, and Applications

Bustard D, Sterritt R (2006) A Requirements Engineering Perspective on Autonomic Systems Development, in Autonomic Computing – Concepts, Infrastructure, and Applications.

Cervenka R, Greenwood D, Trencansky I (2006) The AML Approach to Modeling Autonomic Systems, Whitestein Technologies, Presented at the International Conference on Autonomic and Autonomous Systems (ICAS), July 19-21, 2006, Silicon Valley, USA

Chess D, Hanson J, Kephart J, Whalley I, White S (2006) Dynamic Collaboration in Autonomic Computing, in Autonomic Computing – Concepts, Infrastructure, and Applications

Ganek A (2006) Overview of Autonomic Computing: Origins, Evolution, Direction, in Autonomic Computing – Concepts, Infrastructure, and Applications

Gonzalez T (2007) Handbook of Approximation Algorithms and Metaheuristics. Chapman&Hall/Crc Computer and Information Science Series

Horn P (2001) Senior Vice-President, IBM Research. Autonomic Computing: IBM's Perspective on the State of Information Technology, IBM Research, October 2001

IBM (2006) An Architectural Blueprint for Autonomic Computing, White paper by IBM, June 2006

Kephart J, Chess D (2003) The Vision of Autonomic Computing, Computer, vol. 36, pp. 41-50

Madureira A (2003) Meta-Heuristics Application to Scheduling in Dynamic Environments of Discrete Manufacturing. PhD Dissertation. University of Minho, Portugal. (in portuguese)

Madureira A, Santos F, Pereira I (2008) Self-Managing Agents for Dynamic Scheduling in Manufacturing, GECCO'2008 (Genetic and Evolut. Comput. Conference 2008, Atlanta, Georgia (EUA)

Madureira A, Santos J, Fernandes N, Ramos C (2007) Proposal of a Cooperation Mechanism for Team-Work Based Multi-Agent System in Dynamic Scheduling through Meta-Heuristics. IEEE Intern. Symp. on Assembly and Manufacturing (ISAM07), Ann Arbor (USA), pp. 233-238, ISBN: 1-4244-0563-7

Pinedo M (2008) Planning and Scheduling in Manufacturing and Services, Springer Series in Operations Research and Financial Engineering, Springer

Schwan K, Cooper B, Eisenhauer G, Gavrilovska A, Wolf M, Abbasi H, Agarwala S, Cai Z, Kumar V, Lofstead J, Mansour M, Seshasayee B, Widener P (2006) AutoFlow: Autonomic Information Flows for Critical Information Systems, in Autonomic Computing – Concepts, Infrastructure, and Applications

Siarry P (2008) Advances in Metaheuristics for Hard Optimization. Springer-Verlag

Wellner J and Dilger W (1999) Job shop scheduling with multiagents, in Workshop Planen und Konfigurieren

Wooldridge M (2002) An Introduction to Multiagent Systems, John Wiley and Sons

A sensor classification strategy for robotic manipulators using multidimensional scaling technique

Miguel F. M. Lima [1] **and J. A. Tenreiro Machado** [2]

[1] Dept. of Electrical Engineering, School of Technology, Polytechnic Institute of Viseu, Portugal, *(email: lima@mail.estv.ipv.pt)*

[2] Dept. of Electrical Engineering, Institute of Engineering, Polytechnic Institute of Porto, Portugal, *(email: jtm@isep.ipp.pt)*

Abstract This paper analyzes the signals captured during impacts and vibrations of a mechanical manipulator. To test the impacts, a flexible beam is clamped to the end-effector of a manipulator that is programmed in a way such that the rod moves against a rigid surface. Eighteen signals are captured and theirs correlation are calculated. A sensor classification scheme based on the multidimensional scaling technique is presented.

1 Introduction

In practice the robotic manipulators present some degree of unwanted vibrations. The advent of lightweight arm manipulators mainly in the aerospace industry where weight is an important issue, leads to the problem of intense vibrations. On the other hand, robots interacting with the environment often generate impacts that propagate through the mechanical structure and produce also vibrations. Therefore, the manipulator motion produces vibrations, either from the structural modes or from end-effector impacts. In order to analyze these phenomena a robot signal acquisition system was developed.

Due to the multiplicity of sensors, the data obtained can be redundant because the same type of information may be seen by two or more sensors. Because of the cost of the sensors, this aspect can be considered in order to reduce the cost of the system. On the other hand, the placement of the sensors is an important issue in order to obtain the suitable signals of the vibration phenomenon. Moreover, the study of these issues can help in the design optimization of the acquisition system. In this line of thought a sensor classification scheme is presented.

Several authors have addressed the subject of the sensor classification scheme. White [White 1987] presents a flexible and comprehensive categorizing scheme that is useful for describing and comparing sensors. The author organizes the sensors according to several aspects: measurands, technological aspects, detection

A. Madureira et al. (eds.), *Computational Intelligence for Engineering Systems: Emergent Applications*, Intelligent Systems, Control and Automation: Science and Engineering 46, DOI 10.1007/978-94-007-0093-2_3, © Springer Science+Business Media B.V. 2011

means, conversion phenomena, sensor materials and fields of application. Michahelles and Schiele [Michahelles and Schiele 2003] systematize the use of sensor technology. They identified several dimensions of sensing that represent the sensing goals for physical interaction. A conceptual framework is introduced that allows categorizing existing sensors and evaluates their utility in various applications. This framework not only guides application designers for choosing meaningful sensor subsets, but also can inspire new systems and leads to the evaluation of existing applications.

Today's technology offers a wide variety of sensors. In order to use all the data from the diversity of sensors a framework of integration is needed. Sensor fusion, fuzzy logic, and neural networks are often mentioned when dealing with problem of combing information from several sensors to get a more general picture of a given situation. The study of data fusion has been receiving considerable attention [Esteban et al. 2005, Luo and Kay 1990]. A survey of the sensor fusion techniques for robotics can be found in [Hackett and Shah 1990]. Henderson and Shilcrat [Henderson and Shilcrat 1984] introduced the concept of logic sensor that defines an abstract specification of the sensors to integrate in a multisensor system.

The recent developments of micro electro mechanical sensors (MEMS), with unwired communication capabilities, allow a sensor network with interesting capacity. This technology was adopted in several applications [Arampatzis and Manesis 2005], including robotics. Cheekiralla and Engels [Cheekiralla and Engels 2005] proposed a classification of the unwired sensor networks according to its functionalities and properties.

This paper presents a development of a sensor classification scheme based on the multidimensional scaling technique.

Bearing these ideas in mind, this paper is organized as follows. Section 2 describes briefly the robotic system enhanced with the instrumentation setup. Sections 3 and 4 present some fundamental concepts, and the experimental results, respectively. Finally, section 5 draws the main conclusions and points out future work.

2 Experimental platform

The developed experimental platform has two main parts: the hardware and the software components [Lima et al. 2005]. The hardware architecture is shown in Fig. 3.1. Essentially it is made up of a robot manipulator, a personal computer (PC), and an interface electronic system.

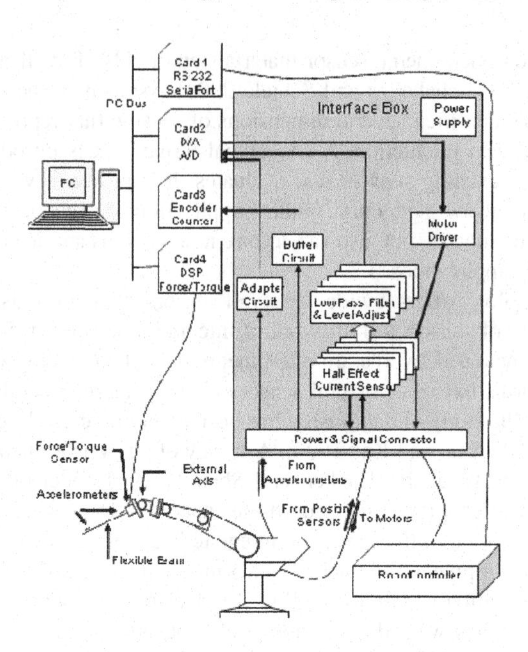

Fig. 3.1 Block diagram of the hardware architecture

The interface box is inserted between the robot arm and the robot controller, in order to acquire the internal robot signals; nevertheless, the interface captures also external signals, such as those arising from accelerometers and force/torque sensors. The modules are made up of electronic cards specifically designed for this work. The function of the modules is to adapt the signals and to isolate galvanically the robot's electronic equipment from the rest of the hardware required by the experiments.

The software package runs in a Pentium 4, 3.0 GHz PC and, from the user's point of view, consists of two applications: the acquisition application and the analysis package. The acquisition application is a real time program for acquiring and recording the robot signals.

After the real time data acquisition, the analysis package processes the data offline in two phases, namely, pre-processing and processing. The preprocessing phase consists of the signal selection in time, and their synchronization and truncation. The processing stage implements several algorithms for signal processing such as the auto and cross correlation, and Fourier transform (FT).

3 Main concepts

This section presents a review of the fundamental concepts involved with Multidimensional scaling (MDS) and metrics in the time domain, namely the correlation.

3.1 Multidimensional scaling

The MDS has its origins in psychometrics and psychophysics where is used as a tool for perceptual and cognitive modeling. From the beginning MDS has been applied in many fields, such as psychology, sociology, anthropology, economy, educational research, etc. In last decades this technique has been applied also in others areas, including computational chemistry [Glunt et al. 1993], machine learning [Tenenbaum et al. 2000], concept maps [Martinez–Torres et al. 2005] and wireless network sensors [Mao and Fidan 2009].

MDS is a generic name for a family of algorithms that construct a configuration of points in a low dimensional space from information about inter-point distances measured in high dimensional space. The new geometrical configuration of points, which preserves the proximities of the high dimensional space, allows gaining insight in the underlying structure of the data and often makes it much easier to understand.

The problem addressed by MDS can be stated as follows: given n items in a m–dimensional space and an $n \times n$ matrix of proximity measures among the items, MDS produces a p-dimensional configuration X, $p \leq m$, representing the items such that the distances among the points in the new space reflect, with some degree of fidelity, the proximities in the data. The proximity measures the (dis)similarities among the items, and, in general, it is a distance measure: the more similar two items are, the smaller their distance is. The Minkowski distance metric provides a general way to specify distance for quantitative data in a multidimensional space:

$$d_{ij} = \left(\sum_{k=1}^{m} w_k \left| x_{ik} - x_{jk} \right|^r \right)^{1/r} \tag{1}$$

where m is the number of dimensions, x_{ik} is the value of dimension k for object i and w_k is a weight. For $w_k = 1$, with $r = 2$, the metric equals the Euclidian distance metric, while $r = 1$ leads to the city-block (or Manhattan) metric. In practice, normally the Euclidian distance metric is used but there are several others definitions that can be applied, including for binary data [Cox and Cox 2001].

Typically MDS is used to transform the data into two or three dimensions, and visualizing the result to uncover hidden structure in the data, but any $p \leq m$ is also possible. A rule of thumb to determine the maximum number of m, is to ensure that there are at least twice as many pairs of items then the number of parameters to be estimated, resulting in $m \geq 4p + 1$ [Carreira-Perpinan 1997]. The geometrical representation obtained with MDS is indeterminate with respect to translation, rotation, and reflection [Fodor 2002].

There are two forms of MDS: metric MDS and nonmetric MDS. The metric MDS uses the actual values of dissimilaries, while nonmetric MDS can use only their ranks. Metric MDS assumes that the dissimilarities δ_{ij} calculated in the original m–dimensional data and distances d_{ij} in the p–dimensional space are related as follows $d_{ij} \approx f(\delta_{ij})$, where f is a continuous monotonic function. Metric (scaling) refers to the type of transformation f of the dissimilarities and its form determines the MDS model. If $d_{ij} = \delta_{ij}$ (it means $f = 1$) and a Euclidian distance metric is used we obtain the classical (metric) MDS.

In metric MDS the dissimilarities between all objects are known numbers and they are approximated by distances. Thus objects are mapped into a low dimensional space, distances are calculated, and compared with the dissimilarities. Then objects are moved in such way that the fit becomes better, until an objective function is minimized. In the context of MDS this objective function is called stress.

In nonmetric MDS, the metric properties of f are relaxed but the rank order of the dissimilarities must be preserved. The transformation function f must obey the monotonicity constraint $\delta_{ij} < \delta_{rs} \Rightarrow f(\delta_{ij}) \leq f(\delta_{rs})$ for all objects. The advantage of nonmetric MDS is that no assumptions need to be made about the underlying transformation function f. Therefore, it can be used in situations that only the rank order of dissimilarities is known (ordinal data). Additionally, it can be used in cases which there are incomplete information. In such cases, the configuration X is constructed from a subset of the distances, and, at the same time, the other (missing) distances are estimated by monotonic regression.

In nonmetric MDS it is assumed that $d_{ij} \approx f(\delta_{ij})$, therefore $f(\delta_{ij})$ are often referred as the disparities [Kruskal and Wish 1978, Martinez and Martinez 2005] in contrast to the original dissimilarities δ_{ij}, on one hand, and the distances d_{ij} of the configuration space, on the other hand. In this context, the disparity is a measure of how well the distance d_{ij} matches the dissimilarity δ_{ij}.

There is no rigorous statistical method to evaluate the quality and the reliability of the results obtained by an MDS analysis. However, there are two methods used often for that purpose: The Shepard plot and the stress. The Shepard plot [Shepard 1962] is a scatterplot of the dissimilarities and disparities against the distances, usually overlaid with a line with a unitary slope. The Shepard plot provides a qualitative evaluation of the goodness of fit, while the stress value gives a quantitative evaluation. Additionally, the stress plotted as a function of dimensionality can be used to estimate the adequate p–dimension. When the curve ceases to decrease significantly we found an "elbow" that may correspond to a substantial improvement in fit.

Beyond the aspects referred before, there are others developments of MDS that includes the replicated MDS and weight MDS. The replicated MDS allows the analysis of several matrices of dissimilarity data simultaneously. The weighted MDS generalizes the distance model as defined in (1).

3.2 *The Correlation coefficient*

Several indices can be used to evaluate the relashionship between the signal, including statistical, entropy and information theory approaches. These metrics are based on a bidimensional probability density function associated with the two signals $x_1(t)$ and $x_2(t)$ acquired in the same time interval and can be calculated according with the expression:

$$P(x_1,x_2)=\frac{\beta(x_1,x_2)}{\iint \beta(x_1,x_2)dx_1dx_2} \tag{2}$$

where β is the bidimensional histogram.

The marginal probability distributions of the signals $x_1(t)$ and $x_2(t)$ are denoted as $P(x_1)$ and $P(x_2)$, respectively. The expected values, $E(x_1)$ and $E(x_2)$, and the variances, $V(x_1)$ and $V(x_2)$, are then easily obtained.

The correlation coefficient R [Orfanidis 1996] is a statistical index that provides a measurement of the similarity between two signals $x_1(t)$ and $x_2(t)$ and is define as

$$R(x_1,x_2)=\frac{E(x_1x_2)-E(x_1)E(x_2)}{\sqrt{V(x_1)V(x_2)}} \tag{3}$$

where $E(x_1x_2)$ is the joint expected value.

4 Experimental results

According to the platform described in section 2 a set of experiments is developed. Based on the signals captured from the robot this section presents several results obtained both in the time and frequency domains.

In the experiments a flexible link is used, consisting of a long and round flexible steel rod clamped to the end-effector of the manipulator. In order to analyze the impact phenomena in different situations two types of beams are adopted. Their physical properties are shown in Table 3.1. The robot motion is programmed in a way such that the rods move against a rigid surface. Fig. 3.2 depicts the robot with the flexible link and the impact surface.

During the motion of the manipulator the clamped rod is moved by the robot against a rigid surface. An impact occurs and several signals are recorded with a sampling frequency of f_s = 500 Hz. The signals come from several sensors, such as accelerometers, force and torque sensor, position encoders, and current sensors.

In order to have a wide set of signals captured during the impact of the rods against the vertical screen thirteen distinct trajectories were defined. Those trajectories are based on several points selected systematically in the workspace of the robot, located on a virtual Cartesian coordinate system (see Fig. 3.3). This coordinate system is completely independent from that used on the measurement system. For each trajectory the motion of the robot begins in one of these points, moves against the surface and returns to the initial point. A parabolic profile was used for the trajectories.

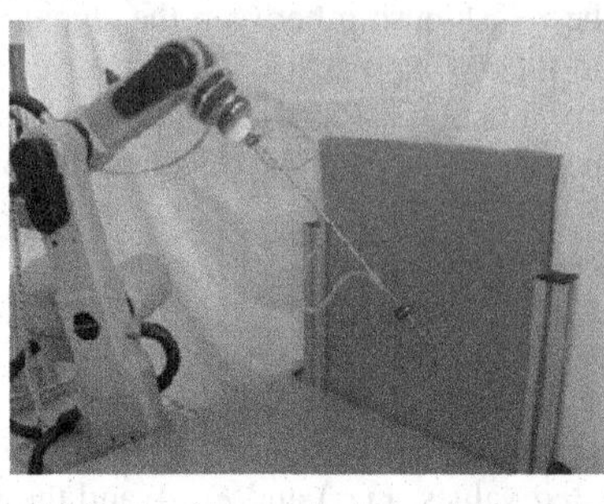

Fig. 3.2 Steel rod impact against a rigid surface

Table 3.1 Physical properties of the flexible beams

Characteristics	Thin rod	Gross rod
Material	Steel	Steel
Density [kg m^{-3}]	4.34×10^3	4.19×10^3
Mass [kg]	0.107	0.195
Length [m]	0.475	0.475
Diameter [m]	5.75×10^{-3}	7.9×10^{-3}

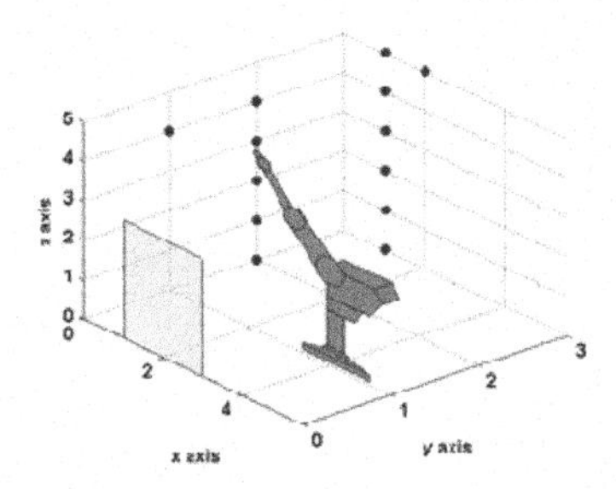

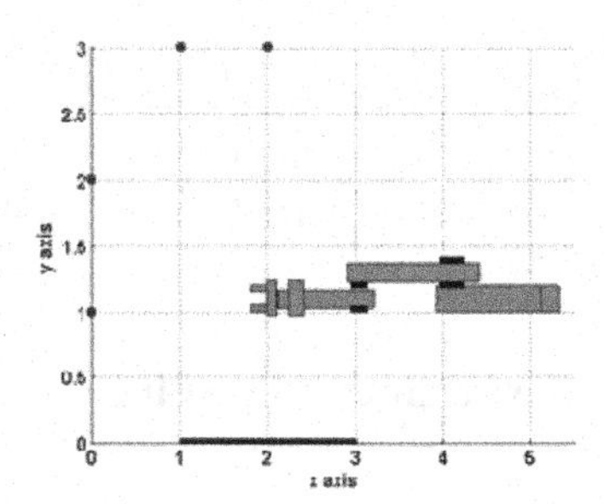

Fig. 3.3 Schematic representation 3D (left) and 2D (right) of the robot and the impact surface on the virtual cartesian coordinate system

4.1 Analysis in the time domain

Fig. 3.4 to 3.7 depict some of the signals corresponding to the cases: (*i*) without impact, (*ii*) with impact of the rod on a gross screen and (*iii*) with impact of the rod on a thin screen, using either the thin, or the gross rod.

In this chapter only the most relevant signals are depicted, namely the forces and moments at the gripper sensor, the electrical currents of the robot's axes motors, and the rod accelerations. The signals present clearly a strong variation at the instant of the impact that occurs, approximately, at $t = 3$ s. Consequently, the effect of the impact forces (Fig. 3.4 left) and moments (Fig. 3.4 right) is reflected in the current required by the robot motors (Fig. 3.6). Moreover, as expected, the amplitudes of forces due to the gross screen (case *ii*) are higher than those corresponding to the thin screen (case *iii*). On the other hand, the forces with the gross rod (Fig. 3.4 right) are higher than those that occur with the thin rod (Fig. 3.4 left). The torques present also an identical behavior in terms of its amplitude variation for the tested conditions (see Fig. 3.5).

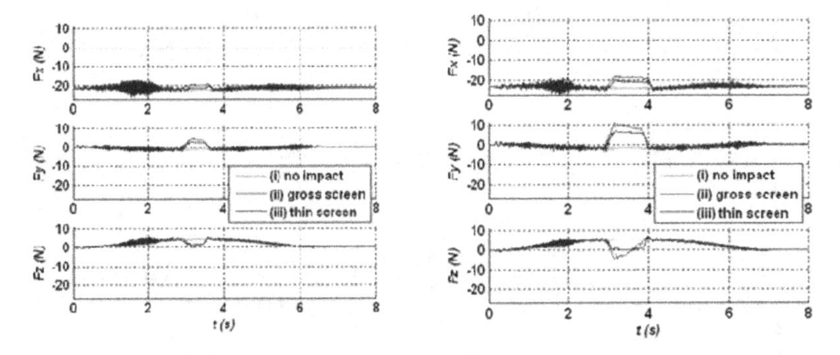

Fig. 3.4 Forces { F_x, F_y, F_z } at the gripper sensor: thin rod (left); gross rod (right)

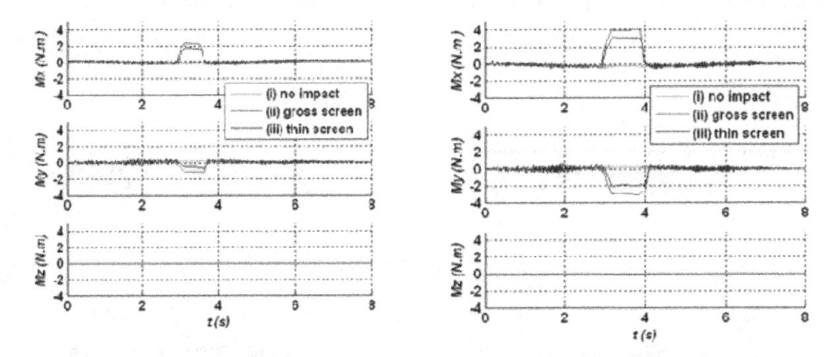

Fig. 3.5 Moments { M_x, M_y, M_z } at the gripper sensor: thin rod (left); gross rod (right)

Fig. 3.7 presents the accelerations at the rod free-end (accelerometer 1), where the impact occurs, and at the rod clamped-end (accelerometer 2). The amplitudes of the accelerometers signals are higher near the rod impact side. Furthermore, the values of the accelerations obtained for the thin rod (Fig. 3.7 left) are higher than those for the gross rod (Fig. 3.7 right), because the thin rod is more flexible.

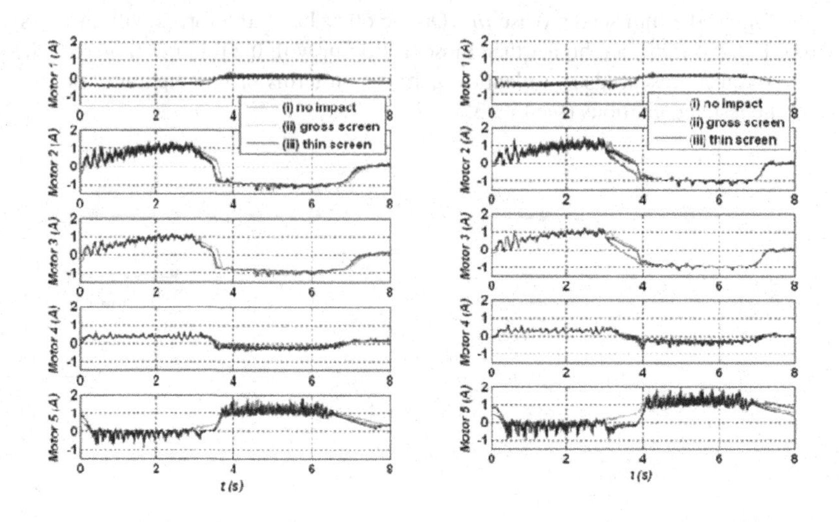

Fig. 3.6 Electrical currents { I_1, I_2, I_3, I_4, I_5 } of the robot's axes motors: thin rod (left); gross rod (right)

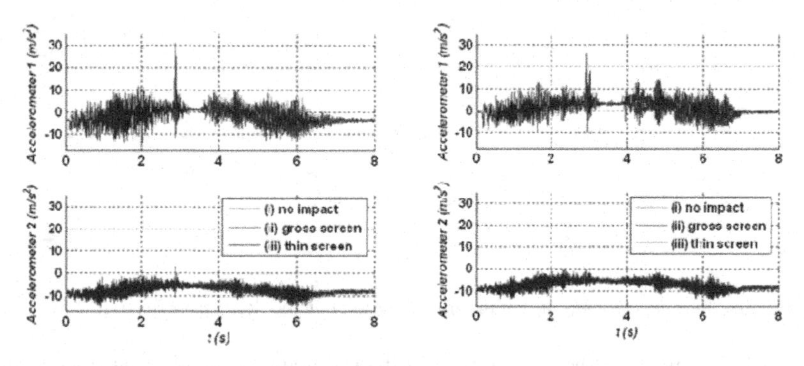

Fig. 3.7 Rod accelerations { A_1, A_2 }: thin rod (left); gross rod (right)

4.2 Sensor classification

Fig. 3.8 shows the squared correlation coefficient R^2 between the signals captured during the same impact trajectory, for an experiment in the case of (*i*) using the gross rod. The results obtain with R^2 are simetric relatively to the main diagonal of the matrix formed by $R^2(x_i,x_j)$, $i = j$, where the metric has a maximum, as expected. To clearly visualize the results only one half is depicted. The correlation between the same families of signals is higher than the correlation between different families. For example, the correlation between the currents and positions is low. The same occurs between the currents and the forces, moments and accelerations. On the other hand, it exists a strong correlation between the positions and the forces, moments and accelerations that depends, as expected, on the trajectory.

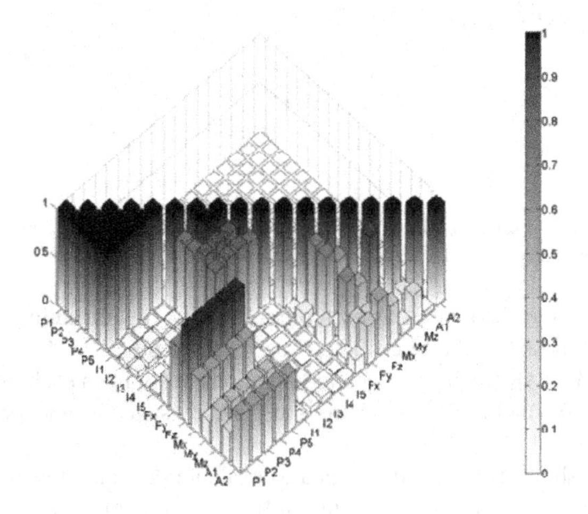

Fig. 3.8 Correlation between the signals {P_n–positions, I_n–electrical currents, F_n–forces, M_n–moments, A_n–accelerations} for the case (*i*) using the gross rod

In order to reveal some hypothetical hidden relationships between the signals the MDS technique is used. Several MDS criteria were tested. The Sammon [Sammon 1969] criterion revealed good results and is adopted in this work. Unlike in usual MDS, this nonlinear mapping technique gives weight to small distances, which helps to detect clusters. In Fig. 3.9 is shown the 2–D (left) and 3–D (right) locus of sensor positioning based on the correlation measure between the signal for the case (*i*) using the gross rod. Three groups of signals can be defined. The ellipses depicted in the chart represent two of these groups. The positions {P_1, P_2, P_3, P_4, P_5} signals are located close to each other. The electrical currents {I_1, I_2, I_3, I_4, I_5} are situated on

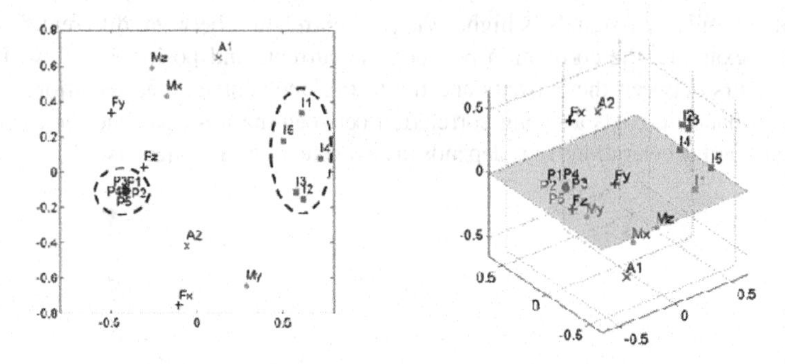

Fig. 3.9 Locus of sensor positioning based on the correlation measure between the signal for the case (*i*) using the gross rod: 2D (left); 3D (right)

the right of the chart and near each other. Finally, the remaining signals form a big group composed by the forces $\{F_x, F_y, F_z\}$, moments $\{M_x, M_y, M_z\}$ and the accelerations $\{A_1, A_2\}$ situated at scattered positions away from each other. A deeper insight into the nature of this feature must be envisaged to understand the behavior of these signals.

Fig. 3.10 shows two tests developed to evaluate the consistency of the results obtained by MDS analysis. The value of the stress function *versus* the dimension is shown in Fig. 3.10 (left), which allows the estimation of the adequate p–dimension. An "elbow" occurs at dimension three for a low value of stress, which corresponds to a substantial improvement in fit. Additionally, the Shepard plot (Fig. 3.10 right) shows the fitting of the 3–D configuration distances to the dissimilarities.

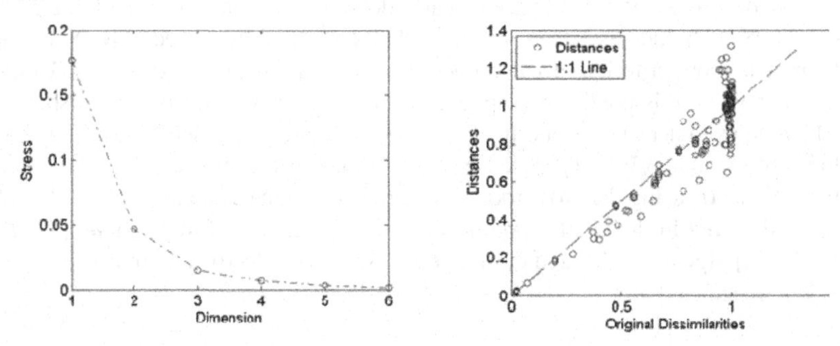

Fig. 3.10 Evaluation of MDS results: Stress test (left); Shepard plot (right)

5 Conclusion

In this paper an experimental study was conducted to investigate several robot signals. A new sensor classification strategy was proposed. The adopted methodology revealed hidden relationships between the robotic signals and leads to arrange them in three groups.

The results merit further investigation as they give rise to new valuable concepts towards instrument control applications. In this line of thought, in future, we plan to pursue several research directions to help us further understand the behavior of the signals. The classification presented was obtained for an experiment corresponding to one trajectory. In future this approach should be applied for all the thirteen trajectories referred before. In this perspective, the replicated MDS technique can be used to analyze simultaneously the respective matrices of dissimilarity. Additionally, others metrics such as the entropy and the mutual information will be used as proximity measures for the MDS technique.

References

Arampatzis T, Manesis S (2005) A Survey of Applications of Wireless Sensors and Wireless Sensor Networks. Proc. IEEE Int. Symp. on Intelligent Control. pp. 719-724

Carreira-Perpinan M (1997) A review of dimension reduction techniques. Technical report CS-96-09, Department of Computer Science, University of Shefield

Cheekiralla S, Engels W (2005) A functional taxonomy of wireless sensor network devices. 2nd International Conference on Broadband Networks IEEE. - Vol. 2. - pp. 949-956. doi:10.1109/ICBN.2005.1589707

Cox, T, Cox M (2001) Multidimensional scaling, 2nd edition, Chapman & Hall/CRC, ISBN 1584880945

Esteban J, Starr A, Willetts R, Hannah P, Bryanston-Cross P (2005) A review of data fusion models and architectures: towards engineering guidelines. Neural Computing & Applications, Springer , London, 14(4):pp. 273–281

Fodor I (2002) A survey of dimension reduction techniques, Technical Report, Center for Applied Scientific Computing, Lawrence Livermore National Laboratory.

Glunt W, Hayden T, Raydan M (1993) Molecular conformation from distance matrices. J. Computational Chemistry, 14, 114-120

Hackett J, Shah M (1990) Multi-sensor fusion: a perspective. Proc. IEEE Int. Conf. on Robotics & Automation, pages 1324–1330

Henderson T, Shilcrat E (1984) E. Logical sensor systems. J. of Robotic Systems. - 2. - Vol. 1. - pp. 169-193

Kruskal J, Wish M (1978) Multidimensional Scaling, Newbury Park, CA: Sage Publications, Inc

Lima M, Machado J, Crisóstomo M (2005) Experimental Set-Up for Vibration and Impact Analysis in Robotics, WSEAS Trans. on Systems, Issue 5, vol. 4, May, pp. 569-576

Luo R, Kay M (1990) A tutorial on multisensor integration and fusion. IEEE 16th Annual Conf. of Industrial Electronics Society, pp. 707–722

Mao Guoqiang, Fidan B (2009) Localization Algorithms and Strategies for Wireless Sensor Networks,Igi-Global, ISBN 978-1-60566-397-5 (ebook)

Martinez W, Martinez A (2005) Exploratory Data Analysis with MATLAB, Chapman & Hall/CRC Press UK, ISBN 1-58488-366-9

Martinez–Torres M, BarreroGarcia F, ToralMarin S, Gallardo S (2005) A Digital Signal Processing Teaching Methodology Using Concept-Mapping Techniques, IEEE Transactions on Education, Volume 48, Issue 3, Aug. 2005 Page(s): 422 – 429 DOI:10.1109/TE.2005. 849737

Michahelles F, Schiele B (2003) Sensing opportunities for physical interaction. Workshop on Physical Interaction of Mobile HCI conference, September, Udine, Italy

Orfanidis S (1996) Optimum Signal Processing. An Introduction. 2nd Edition, Prentice-Hall, Englewood Cliffs, NJ

Sammon J (1969) A nonlinear mapping for data structure analysis. IEEE Trans. Computers, C–18(5): 401–409

Shepard R (1962) The analysis of proximities: multidimensional scaling with an unknown distance function, I and II, Psychometrika, 27, pp. 219-246 and pp. 219-246

Tenenbaum J, de Silva V, Langford, J (2000) A global geometric framework for nonlinear dimensionality reduction. Science, 290(5500), 2319-2323

White R (1987) A sensor classification scheme. IEEE Trans. on Ultrasonics, Ferroelectrics and Frequency Control, 34(2):124–126

Collective-Intelligence and Decision-Making

Paulo Trigo[1] **and Helder Coelho**[2]

[1] ISEL, Lisbon, Portugal, *(email: ptrigo@deetc.isel.ipl.pt)*

[2] FCUL, Lisbon, Portugal *(email: hcoelho@di.fc.ul.pt)*

Abstract Decision-making, while performed by humans, is also expected to be found in most (artificial) intelligent systems. Usually, the cognitive research assumption is that the individual is the correct unit for the analysis of (human) intelligence. Yet, the multi-agent assumption is that of a society of interacting individuals (agency) that collectively supersedes individual capabilities. Therefore, the entire society of agents is, itself, an additional locus for the analysis of this collective-intelligence. In this paper we propose models that explore the agent-agency mutual influence from the decision-making perspective. We outline three case study scenarios for the models' experimental evaluation: i) large-scale disasters, ii) electricity markets, and iii) Web-empowered knowledge and social connectivity. The scenario-driven evaluations are being used to guide our research efforts.

1 Introduction

"What do you think about our course in this institution? — the younger student asks two older students while looking at a professor that roves nearby. "This daily, face-to-face, scene enfolds individuals in two major ways: i) throughout knowledge sharing, and ii) by refining (clarifying) their social network connectivity. Now, imagine that same student exploring the Internet, e.g., via search engines, electronic encyclopedias or Weblogs (blogs), looking for answers and also providing his own opinion on that same question. In both scenes (face-to-face and Internet) each individual's decision-making influences others while also being influenced by them. The resulting overall knowledge and group reasoning capability supersedes individuals and frames a collective-intelligence.

There are several perspectives on the collective-intelligence concept. A behavioral approach considers collective intelligence as "consisting of a large number of quasi-independent, stochastic agents, interacting locally both among themselves and with an active environment, in the absence of hierarchical organization, and yet capable of adaptive behavior" [Sulis 1997]. The organizational perspective of collective-intelligence is that of "groups of individuals doing things collectively that seem intelligent" [Malone et al. 2009]. The different collective-intelligence

A. Madureira et al. (eds.), *Computational Intelligence for Engineering Systems: Emergent Applications*, Intelligent Systems, Control and Automation: Science and Engineering 46,
DOI 10.1007/978-94-007-0093-2_4, © Springer Science + Business Media B.V. 2011

perspectives contribute to formulate a basic research question: "How can people and computers be connected so that – collectively – they act more intelligently than any individuals, groups, or computers have ever done before?" [MIT, 2009]. This question brings humans and computers as partners mutually influencing each other; it is not just the computer that exhibits intelligent capabilities but the blending of computers with humans that aims to exceed each other's capabilities. The modern "computer partnership" exposes such "blending" of computers and humans thus being the major novelty and driving force for the collective-intelligence research field.

In this paper we describe our research proposals on the multi-agent decision-making field, mainly focused on domains that exhibit uncertain causal effects. Section 2 describes the essentials of our proposal to decision-making with multiple simultaneous goals from an institutional agent's perspective. In section 3 we extend the previous scope and formulate a model that accounts for the separation of concerns between collective and individual motivations, thus framing a 2-strata decision model, collective 'versus' individual (CvI) where the definition of inter-stratum relations enables each agent to choose at which stratum each decision should be taken. Section 4 applies a simplified CvI model to the electricity market domain with the multi-agent simulator TEMMAS (The Electricity Market Multi-agent Simulator). The last proposal is presented in section 5 with our preliminary work to exploit the cross cutting between the agent-based system modeling and the Web-based methodologies given the emphasis on the emerging ideas (and tools) that populate the Semantic Web (SW) field. Institutional agents can be designed to take advantage of the collective-knowledge-representation perspective of the Internet space and to foster the early adoption, by the institutions, of the SW concepts and technologies, thus expanding the organizations' collective-intelligence.

2 Multiple simultaneous goals and uncertain causality

The mitigation of a large-scale disaster, caused either by a natural or a technological phenomenon (e.g., an earthquake or a terrorist incident), gives rise to multiple simultaneous goals that demand the immediate response of a finite set of specialized agents. A rational agent must evaluate multiple and simultaneous perceived damages, account for the chance of mitigating each damage and establish a preferences relation among goals. Although rational behavior stands on the capability to establish preferences among simultaneous goals, current belief-desire-intention (BDI) mental-state approaches [Raoand Georgeff 1995, Wooldridge 2000] (widely used as reasoning agents' foundations) do not provide a theoretical or architectural framework for predicting how goals interact and how an agent decides which goals to pursue. When faced with multiple simultaneous goals, the BDI's intention selection process (decision) follows heuristic approaches, usually coded by a human designer [Pokahr et al. 2005]. Additionally, BDI models find it difficult to deal with uncertainty, hence hybrid models have been proposed combining BDI and Markov decision process (MDP) approaches [Simariand Parsons 2006, Trigo and Coelho 2007]; however, hybrid models usually assume that the goal has already been chosen and tackle the stochastic planning problem (goal achievement).

In this work we take the decision-theoretic notion of rationality to estimate the importance of goals and to establish a preferences relation among multiple goals. We propose a model that allows agent developers to design the relationships between perceived (certain) and uncertain aspects of the world in an easy and intuitive manner. The design is founded on the influence diagram [Howard and Matheson 1984] (ID) framework that combines uncertain beliefs and the expected gain of decisions.

The proposal's practical usefulness is experimentally explored in a fire fighting scenario in the RoboCupRescue [Kitano and Tadokoro 2001] domain. The decision model in corporate general fire fighting principles in a way that considerably simplifies the specification of a preferences relation among goals. Despite such simplification, the attained results are consistent with the initial fire fighting principles. For the comprehensive description of the proposal we refer to [Trigo and Coelho 2008a].

2.1 The preferences model and the causal effect pattern

The premise of the preferences model is that the relation among simultaneous goals follows from the expected utility of the available decisions. The expected utility of any available decision combines: i) the observed state value, and ii) the likelihood of success of that decision. Given a set of available decisions, $\mathscr{D}$, a set of states, $\mathscr{S}$, an utility function, $u : \mathscr{S} \rightarrow \mathrm{R}$, and the probability, $P(\ s \mid d\ )$, to achieve state $s \in \mathscr{S}$ after decision $d \in \mathscr{D}$, the expected utility, $eu : \mathscr{D} \rightarrow \mathrm{R}$, of decision-making is described by, $eu\ (D = d\ ) = \sum_{s \in \mathscr{S}} P\ (\ s \mid D = d\ )\ u(\ s\ )$, where D is a variable that holds an available decision.

Give nany goal there are always two available decisions: i) pursue the goal, or ii) ignore the goal. Thus, $\mathscr{D} = \{\texttt{yes}, \texttt{no}\}$, is such that $D_g = \texttt{yes}$ and $D_g = \texttt{no}$ represent, respectively, the decision to pursue or to ignore the goal $g \in \mathscr{G}$. The utility of a goal, g, measures the importance, assigned by the agent, to the goal g. The "importance" is a criterion related to a valuation in terms of benefits and costs an agent has of a mental state situation [Corrêa and Coelho, 2004]. The goal achievement payoff is estimated by the difference between the expected utility on pursuing and ignoring that goal. Thus, the goal utility function, $u_{\mathscr{G}}$, for each $g \in \mathscr{G}$, is defined by, $u_{\mathscr{G}}(g) = eu\ (D_g = \texttt{yes}) - eu\ (D_g = \texttt{no})$. The agent prefers goal $g1$ over $g2$ (i.e., $g1 > g2$) if g1 offers higher payoff and when even, prefers the goal that, when achieved, gives higher expected advantage (i.e., higher $eu\ (D_g =$

yes)). In sight of equality the agent is indifferent between goals(i.e., $g1 \sim g2$), thus taking, for instance, an exploratory decision.

The causal effects (consequences) of each decision are unknown, therefore our aim is to choose the decision alternative (goal) that minimizes the eventual disadvantageous consequences of such decision. The influence diagram (ID) framework combines uncertain beliefs to compute the expect edutility of decisions, thus rationality is a matter of choosing the alternative that leads to the highest expected utility, given the evidence of available information. The ID extends the Bayesian's network chance nodes with two additional nodes: decisions and utilities, and two additional arcs: influences and informational. We propose a set of (nine) guidelines, to build an ID pattern to describe the multiple and simultaneous goals decision problem. Fig 4.1 illustrates the pattern using the regular ID symbols; circle is a chance node, rectangle is a decision node and the lozenge is an utility node.

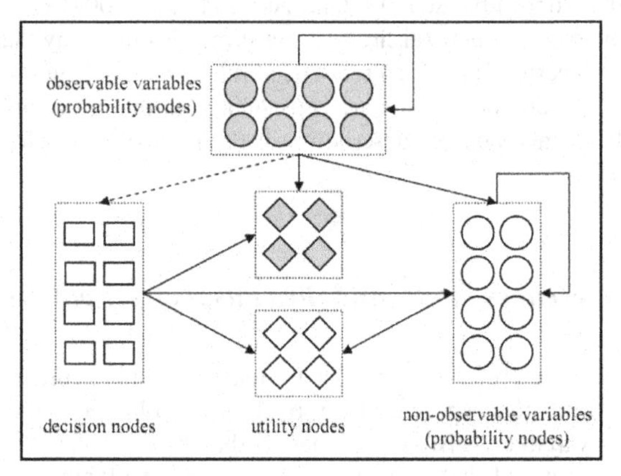

Fig. 4.1 The influence diagram (ID) pattern (the sets are represented by dotted rectangles; the gray elements refer to observable information; the dotted arcs are informational arcs and the remaining are conditional arcs).

The gray filling (cf. figure 4.1) has a special meaning: i) the gray chance node indicates information availability, i.e., an observable variable, and ii) the gray utility node indicates a dependency from a gray chance node, i.e., the utility of some observable variables. The sets of nodes with similar characteristics are aggregated by a dotted rectangle. The arcs connect sets of nodes (instead of individual nodes), therefore attaining an ID pattern, i.e., a template from which to build several different instances with the same overall structure; we next describe the ID pattern usage.

2.2 The experimental scenario (multiple simultaneous goals)

We used the RoboCupRescue environment to devised a disaster scenario that evolves at the Nagata ward in Kobe, Japan. Two buildings, B_1 and B_2, not far from each other (about 90 meters) catch a fire. The B_1 is relatively small and is located near Kobe's harbor, in a low density neighborhood. The B_2 is of medium size and it is highly surrounded by other buildings. As time passes, the fires' intensity increase so a close neighbor is also liable to catch a fire. Fig. 4.2 shows the disaster scenario; each opaque rectangle is a building and a small circle is positioned over B_1 and B_2. The two larger filmy squares define the neighborhood border of B_1 and B_2 within a d distance (in meters).

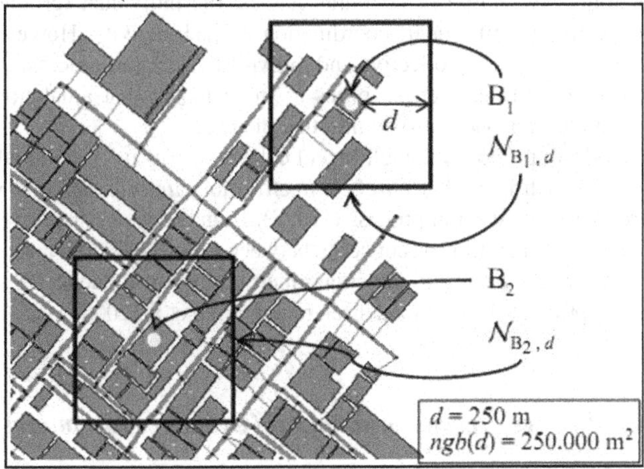

Fig. 4.2 Fire scenario in buildings labeled B_1 and B_2 (the set of buildings that are contained within each building's neighborhood, $ngb(d)$, is represented by $\mathcal{N}_{Bi,d}$).

To simplify we assume that: i) buildings use identical construction materials, ii) buildings are residential (neither offices nor industries inside the buildings), and iii) there are no civilians, caught by fires, inside the buildings. In order to apply the ID pattern (cf. fig. 4.1) to the illustrative cenario (cf. fig.4.2)we formulated, for each building, the observable and non-observable variables along with the utility nodes which followed three general fire attack strategies that, although intuitive, were acquired after the RoboCupRescue experimentation.

2.3 Results and prospects after this work

Practical experiences indicate that the ID pattern considerably simplifies the specification of a decision model (in the RoboCupRescue domain) and enabled to established a preferences order among goals that is consistent with the initial, domain expert, very general strategies. This work represents the ongoing steps in a line of research that aims to develop decision making agents that inhabit complex environments (e.g., the RoboCupRescue). Agter this work we applied the preferences model to the problem of coordinating teamwork (re)formation [Trigo and Coelho 2007] from a centralized perspective.

3 Decisions with collective and individual motivations

The agents that cooperate to mitigate the effects of a large-scale disaster take decisions that follow two large behavioural classes: the individual (ground) activity and the collective (institutional) coordination of such activity. However, despite the intuition on a 2-strata (collective and individual) decision process, research on multi-agent coordination often proposes a single model that amalgamates those strata and searches for optimality within that model.

In this work we propose the multi-agent collective 'versus' individual (CvI) decision model, which is neither *purely collective* nor *purely individual*, is founded on the semi-Markov decision process (SMDP) frame work and is designed to explore the explicit separation of concerns between both (collective and individual) decision strata while aiming to conciliate their reciprocal influence. The comprehensive description of the CvI model refers to [Trigo et al. 2006].

3.1 The collective 'versus' individual (CvI) decision model

The premise of the CvI decision model is that the individual choice coexists with the collective choice and that coordinated behaviour happens (is learned) from the prolonged relation (intime) of the choices exercised at both of those strata (individual and collective). Additionally, coordination is exercised on highlevel *cooperation tasks*, represented within an hierarchical task organization. The tasks' hierarchy is founded on the framework of *Options* [Suttonetal. 1999], which extends the MDP theory to include *temporally abstract actions*, i.e., variable time duration tasks, whose execution resorts to a subset of primitive actions.

3.1.1 The CvI collective and individual strata

The individual stratum is simply a set of agents, [illegible], each agent, $j \in$ [illegible], with its capabilities described as an hierarchy of options. The CvI model admits agent heterogeneity (diverse option hierarchies), as long as all hierarchies have the same number of levels (depth), i.e., a similar temporal abstraction is used to design all hierarchies.

The collective stratum consists of a single agent (e.g. an institutional agent) that represents the whole set of individual stratum agents. The collective stratum agent cannot act on its own; its actions must be materialized through the individual stratum agents. The purpose of the collective stratum is to coordinate the individual stratum. Formally, at the collective stratum, each action is defined as a *collective option*, $o_o = (\mathcal{I}_o, \pi_o, \beta_o)$, where $o = (o^1, \ldots, o^{|⍰|})$ represents the simultaneous execution of option $o_j \equiv (\mathcal{I}^j, \pi^j, \beta^j)$ by each agent $j \in$ ⍰. The set of agents, ⍰, defines an option space, $\mathcal{O} \subseteq \mathcal{O}^1 \times \ldots \times \mathcal{O}^{|⍰|}$, where $\mathcal{O}^j$ is the set of agent j options and each $o_o \in \mathcal{O}$ is a collective option. The $\mathcal{O}$ decomposes into $\mathcal{O}^d$ disjoint subsets, each containing only the collective options available at the, d, hierarchical level, where $0 < d \leq D-1$ and level -0 is the hierarchy root, at which there are no options to choose from, and level -D is the hierarchy depth. A level d policy, π_d, is implicitly defined by the SMDP $\mathcal{M}_d$ with state set $\mathcal{S}$ and action set $\mathcal{O}_d$. The $\mathcal{M}_d$ solution is the optimal way to choose the level d individual policies which, in the long run, gathers the highest collective reward.

3.1.2 The CvI structure and dynamics

Fig. 4.3 illustrates the CvI structure where the individual stratum (each *agent* ^{j}) is a 3-level task hierarchy and thus the collective stratum is a 2-level hierarchy; at each level, the set of diamond ended arcs, links the collective option to each of its individual policies.

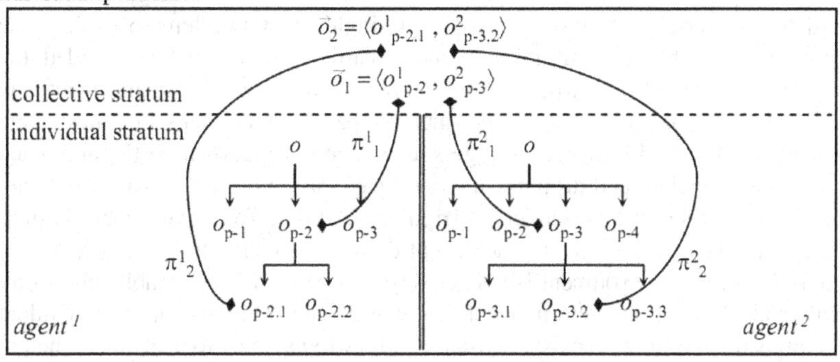

Fig. 4.3 The structure of the CvI decision model and links between strata (super script j refers to agent j; subscripts k and p-k refer to k hierarchical level and k tree path).

Fig. 4.4 illustrates the CvI dynamics; at each decision epoch, *agent* ^{j} gets the partial perception, ω^j, and decide-who-decides (*d-w-d*), i.e., the *agent* ^{j} either: i) chooses an option $o^j \in \mathcal{O}^j$, or ii) requests, the collective stratum, for a decision. The collective stratum always replies with an option, o^j, decision. The *d-w-d* process represents the importance that an agent credits to each stratum motivation, which is materialized as the ratio between, the maximum expected benefit in choosing a collective and an individual decision. The expected benefit is given, at each hierarchical level-d, by the value functions of the corresponding SMDP $\mathcal{M}_d$. A threshold, $\kappa \in [0, 1]$, supports the focus grade between collective and individual strata. Such regulatory mechanism enables the (human) designer to specify diverse social attitudes: ranging from common-good ($\kappa = 0$) to self-interested ($\kappa = 1$) agents.

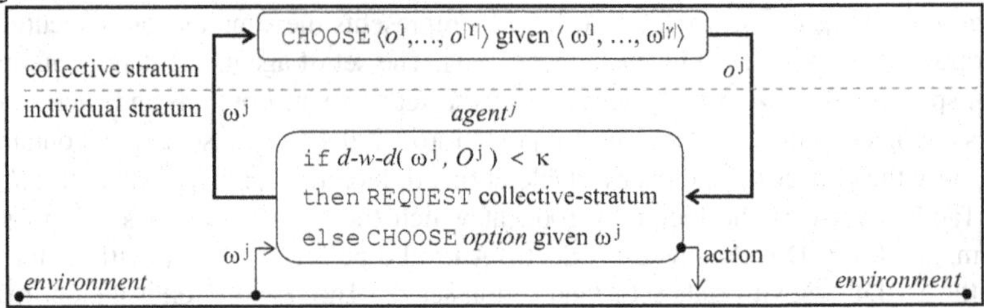

Fig. 4.4 The CvI decide-who-decides (d-w-d) process ($o \equiv$ option; $\mathcal{O} \equiv$ option set; $\omega \equiv$ partial observation; superscript j refers to *agent*^{j}).

3.2 The experimental scenario (ambulances and injured civilians)

We implemented the CvI decision model and tested it in a multi-agent ambulance environment: "a maze-like grid in habited by ambulances (ortaxis), in jured civilians (or passengers) and hospitals (or sites)". The problem was originally formulated by Dietterich [Dietterich 2000] as a single-agent problem to explore the properties of a hierarchical reinforcement learning method. We extended the original problem to the multi-agent environment as follows: "the environment is in habited by multiple taxis (the agents), multiple passengers and a set of sites; a site may be the origin of several passengers, each one with its own destination; each taxi is able to pickup and drop down several passengers; each taxi may simultaneously transport several passengers" [Trigo et al. 2006]. The environment is individually partially observable as each agent does not perceive the other agents' locations; also, the environment is collectively observable as the combination of all individual observations determines a sole world state. The goal of the individual stratum is to learn how to execute tasks (e.g. how to navigate to a site and when to pick up a passenger). The goal of the collective stratum is to learn to coordinate those individual tasks as to minimize the resources (time) to satisfy the passengers' needs. The learning of the policy at the collective stratum occurs simultaneously with the learning of each agent's policy at the individual stratum.

3.3 Results and prospects after this work

The formulation of the CvI multi level hierarchical decision model and the definition of an inter-strata regulatory mechanism enable us to show experimentally how to explore the individual policy space in order to decrease the complexity of learning a coordination policy in a partially observable setting. The 2-strata approach augments the design flexibility in two ways: i) makes it possible to specify individual task hierarchies that are not necessarily equal, therefore allowing for agents' heterogeneity, and ii) enables to configure different architectures (e.g. centralized or decentralized) depending on the information exchange between collective and individual layers.

Our contribution was extended with the formulation of the CvI-JI hybrid model [Trigo and Coelho 2010]. We have identified a series of relations between the CvI decision-theoretic approach and the joint-intentions (JI) [Cohen and Levesque 1991] mental-state based reasoning. The CvI was extended by exploring the algorithmic aspects of the CvI-JI integration. Such integration represents our novel contribution to a multi-agent hybrid decision model within a reinforcement learning framework.

4 Decision-making for electricity markets

The start-up of nation-wide electric markets, along with its recent expansionto intercountry markets, aims at providing a competitive electricity service to consumers. This newly deregulated electricity market organization calls for an increasing (human) decision-making responsibility in order to settle the energy assets' trading strategies. The growing number of interactions among market participants and their mutual-influencing are usually described by game theoretic approaches which are based on the determination of equilibrium points against which to compare the actual market performance [Berryetal. 1999, Gabrieletal. 2004]. However, those approaches find it difficult to incorporate the ability of market participants to repeatedly probe markets and adapt their strategies. As an alternative to the equilibrium approaches, the multi-agent based simulation (MABS) comes forth as being particulary well fitted to analyze dynamic and adaptive systems with complex interactions among constituents [Schuster and Gilbert, 2004, Helleboogh et al. 2007].

In this work we propose a MABS to capture the behaviour of the electricity market and to build TEMMAS (The Electricity Market Multi-agent Simulator), which incorporates the operation of multiple generator company (*GenCo*) operators, each with distinct power generating units (*GenUnit*), and a market operator (Pool) which computes the hourly market price (driven by the electricity demand). We used TEMMAS to construct an agency for the Iberian (Portugal and Spain) Electricity Market (MIBEL – "Mercado Ibérico de Electricidade"). The agency follows the market organizational structure and each agent acts (bid,buy,sell) according to its role in the market. The simulation explores the relation between the production capacity and the search for competitive bidding strategies. Results show a coherent market behaviour as price reductions from competiti on favour consumers and harm producers that do not adapt their electricity bids. The comprehensive description of TEMMAS refers to [Trigo and Coelho 2008b, Trigo et al. 2010].

4.1 TEMMAS agency design

Within the current design model of TEMMAS the electricity asset is traded through a spot market (no bilateral agreements), which is operated via a *Pool* institutional power entity. Each generator company, *GenCo*, submits (to *Pool*) how much energy, each of its generating unit, $GenUnit_{GenCo}$, is willing to produce and at what price. The bidding procedure conforms to the so-called "block bids" approach [OMIP,], where a block represents a quantity of energy being bided for a certain price; also, *GenCos* are not allowed to bid higher than a predefined price ceiling. Thus, the market supply essential measurable aspects are the energy price, quantity and marginal production cost. The consumer side of the market is mainly described by the quantity of demanded energy; we assume that there is no price elasticity of demand(i.e., no demand-side market bidding). The Pool is a reactive agent that always applies the same predefined auction rules in order to determine the market price and hence the block bids that clear the market. Each $\mathcal{E}_{\text{GenUnitGenCo}}$ represents the *GenCo's* set of available resources. The $\mathcal{E}_{\text{GenCo}}$ contains the decision-making agents.

Each generator company defines the bidding strategy for each of its generating units. We designed two types of strategies: a) the basic-adjustment, that chooses among a set of basic rigid options, and b) the *heuristic-adjustment*, that selects and follows a predefined well-known heuristic. The strategies correspond to the *GenCo* agent's primary actions. The *GenCo* has a set, $\mathcal{E}_{GenUnitGenCo}$, of generating units and, at each decision-epoch, it decides the strategy to apply to each generating unit, thus choosing a vector of strategies, $\overrightarrow{stg}$ where the i th vector's component refers to the $GenUnit^{i}_{GenCo}$ generating unit. The *GenCo*'s perceived market share, *mShare*, is used to characterize the agent internal memory so its state space is given by *mShare* $\in$ [0..100]. Each *GenCo* is a MDP decision-making agent such that the decision process period represents a daily market. At each decision-epoch each agent computes its daily profit (that is regarded as an internal reward function) and the *Pool* agent receives all the *GenCos*'s block bids for the 24 daily hours and settles the hourly market price by matching offers in a classic supply and demand equilibrium price (we assume a hourly constant demand).

Fig 4.5 shows TEMMAS agents along with the major inter-agent communication paths; at the top there is an user interface to specify each of the resources' and agents' configurable parameters. The implementation of the TEMMAS architecture followed the INGENIAS [Gómez-Sanz et al., 2008] methodology and used its supporting development platform.

4.2 The experimental scenario (Iberian electricity market)

We used TEMMAS to build a specific electric market simulation model taken from the Iberian Electricity Market (MIBEL– "Mercado Ibérico de Electricidade") with Portuguese (e.g., EDP - "Electricidade de Portugal", "Turbogás", "TejoEnergia") and Spanish (e.g., "Endesa", "Iberdrola", "UnionFenosa", "Hidro-Cantábrico", "Viesgo", "BasNatural", "Elcogás") generator companies. Regarding the total electricity capacity installed the Iberian market is composed of a major player (Spain) and a minor player (Portugal). Fig 4.5 uses INGENIAS not ationto depict the hierarchical structure of the electricity market. We considered three types of generating

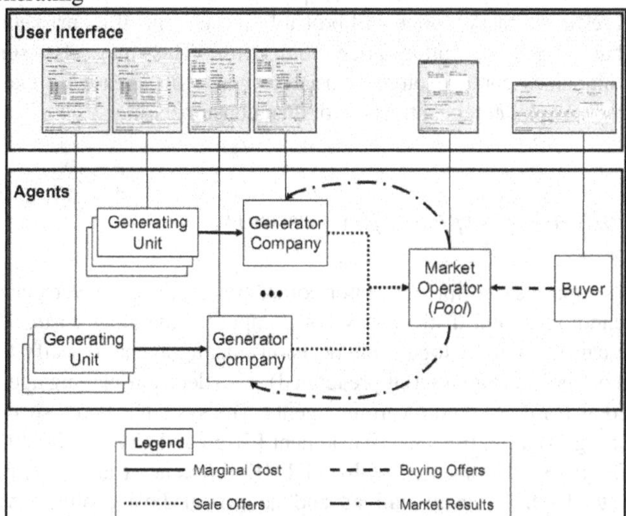

Fig. 4.5 The TEMMAS architecture and the configurable parameters.

units: i) coal plant, CO, to provide the base load demand, ii) combined cycle plant, CC, to cover intermediate load, and iii) gas turbine, GT, to cover peaking loads.

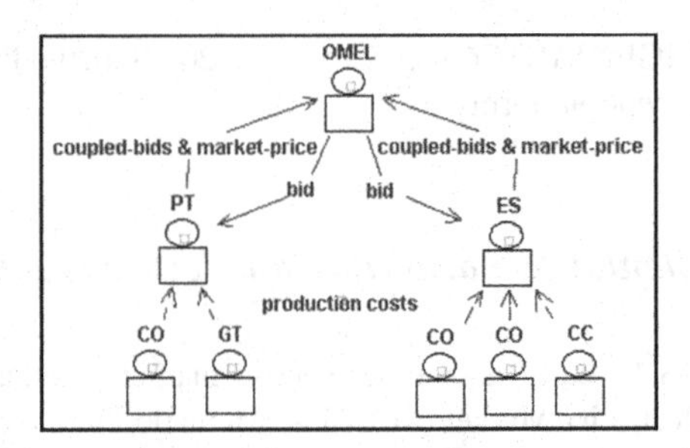

Fig. 4.6 An illustrative TEMMAS formulation (with INGENIAS notation); PT and ES are, respectively, Portugal and Spain; OMEL represents the Iberian market operator

The simulation shows how a one-company market will find its way to sell all its electricity at the price of its most expensive energy (e.g., sell the cheapest CO at the price of the expensive CC and GT). The results for the two-company market show that a big company, capable of supplying the whole demand, with a strategy to always settle the market price will probably loose some of its market share for a smaller but aggressive cutting-price competitor. This may also suggest that smaller companies could explore alternative ways to increase its market share, such as attractive bilateral contracts with final consumers.

4.3 Results and prospects after this work

TEMMAS supports multiple generator companies, multiple power plant with different technologies and distinct ways of computing marginal costs. The user is given the capability to configure the decision-making agents (specify the bidding strategies or choose from a set of predefined); also decision-making agents may be configured as learning or non-learning agents. The system is being specified (from the beginning) as a multi-agent environment [Trigo and Coelho 2008b], modelled with INGENIAS, implemented with JADE [Bellifemine et al. 2007] (using "INGENIAS to JADE" transformations and additional Java coding) and with R-Project [Venables and Smith 2009] for the statistical computations and graphics' generation. Although TEMMAS is currently in a preliminary stage its architectural and design decisions are strongly founded in the MABS field and the initial results are an incentive to further extend the already implemented system. We intend to extend TEMMAS taking into account the particularities of the MIBEL market(e.g., the network congestion and its relation with the market-splitting procedure).

5 Agent inferencing meets the Semantic Web

Agent inferencing over Semantic Web (SW) descriptions gives organizations the opportunity to use the Internet as a "huge" collective-knowledge-representation plat-form. The meaning of this is clear: "there is a next ended spectrum of applications and technologies for storing, searching and retrieving, data, information and knowledge". The progress towards the SW is clear from the standardization of languages and the increasing maturity of related models and tools. The SPARQL (SPARQL Protocol and RDF Query Language) enables to extract the RDF (Resource Description Language) explicitly stored information while the SW rules (or semantic constraints) claim for ways of extracting the implicitly represented information.

We propose to combine the SPARQL expressiveness with the RDF-triple uniform description model in order to transform implicit knowledge in to explicitly stored information. The practical stance is that the currently available SPARQL specifications [Beckett and Broekstra, 2008, Prud'hommeaux and Seaborne, 2008] do not support RDFS entailment, so we combine SPARQL and RDF specifications [Manola and Miller, 2004] to build an inferencing scenario driven by the SW.

The SPARQL language queries the information explicitly represented within an RDF graph. The implicit RDF triples are the ones that may be inferred from the rules that characterize each domain interpretation. The inference process considers the following typical representative of the rule format: ($\alpha \Rightarrow \beta$) where α is the rule antecedent for mulated as a conjunction of several RDF triples and β is the rule consequent formulated as a single RDF triple. The inference process expands a graph $\mathcal{G}$ into a graph $\mathcal{G}$ which explicitly includes the triples that, given rule ($\alpha \Rightarrow \beta$),only existed implicitly in $\mathcal{G}$. The comprehensive description of this proposal refers to [Trigo and Coelho, 2009].

5.1 The experimental scenario (Fire-Brigade decision-making)

The institutions related with fire-brigade domain can take advantage of SW early adoption. Those institutions rely on diverse sources of information thus depending on (automated) mechanisms for concepts' interpretation, classification and integration. For example, the Portuguese SNBPC (Serviço Nacional de Bombeiros e Protecção Civil – "National Fire Service and Civil Defense") must interpret (and-integrate) the information provided by the set of fire-brigade command centers to plan and coordinate the huge amount of (annual) florestal fires; automated knowledge-handling would certainly be a helper "to follow the ambition on reducing" the (typical) Portuguese post-summer annual landscape of grievously huge burnt areas.

The fire-brigade domain has several fire-brigade centers each liable for a region and its set of smaller places (e.g., house blocks in a town). Fires are classified by their priority and several strategies have been devised for scenarios with several simultaneous fires [Trigo and Coelho 2008a] while the decision –making process follows the hierarchical relations among agents [Trigo and Coelho 2010]. Fig. 4.7 shows the basic RDF description of a fire-brigade domain. Each arc is a *property* and each oval node is a *resource*, the node *b* is an anonymous *resource* (i.e., a blank node), the rectangle node *v* is a literal and the qualifier *D* is the domain namespace.

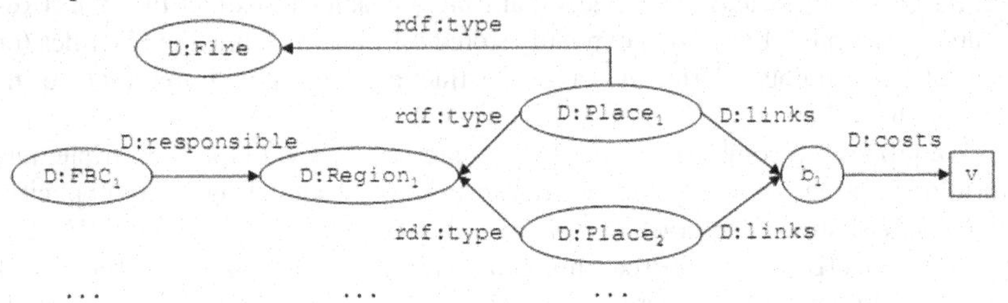

Fig. 4.7 A RDF basic description of fire-brigade domain.

The agent designer uses the RDF framework to describe the domain *resources* and *properties* and then he defines the semantic rules. When the agent receives a fire report the inference process is executed to determine the region of the fire and thus the fire-brigade center that has the responsibility of attacking the fire.

5.2 Results and prospects after this work

This is a preliminary work that is currently being developed in order to exploit the crosscutting between the agent-based system modelling and the Web-based methodologies given the emphasis on the emerging ideas (andtools) that populate the SW field. Institutional agents (e.g., fire-brigade) can be designed to take advantage of the collective-knowledge-representation perspective of the Internet space and to foster the early adoption, by the institutions, of the SW concepts and technologies.

6 Conclusions

This paper describes our research effort along the lines of individual and collective autonomous decision-making. We have designed and implemented agents that decide and act collectively in (simulated) large scale desastres, energy markets and in the semantic Web space. Our research effort have been strongly impelled by the questionon "How to capture the dynamics of the agent-agency mutual influencing indecision-making processes?" The collective 'versus' individual (CvI) modelis our most thorough proposal but humans are not regarded as full participants in the decision-making processes, i.e., the agency excludes humans (thus being purely virtual). The collective-intelligence perspective of "blending" humans and computers frees agents from the responsibility of substituting humanprocesses. Both humans and computers regard the other as an appendix of itself. We intend to research new ways of augmenting our decision-making models with this "blending" modern way of looking at the collective-intelligence phenomenon.

References

Beckett D, Broekstra J (2008) SPARQL query results XML format. W3C Recommendation

Bellifemine F, Caire G, Greenwood D (2007) Developing Multi-Agent Systems with JADE. Wiley Series in Agent Technology. Wiley

Berry C, Hobbs B, Meroney W, O'Neill R, Jr W (1999) Understanding how market power can arise in network competition: a game theoretic approach. Utilities Policy, 8(3): 139–158

Cohen P, Levesque H (1991) Team work. Noûs, Special Issue on Cognitive Science and Artificial Intelligence, 25(4):487–512

Corrêa M, Coelho H (2004) Collective mental states in extended mental states framework. In Proceedings of the IV International Conference on Collective Intentionality, Certosa di Pontignano, Siena, Italy

Dietterich T (2000) Hierarchical reinforcement learning with the MAXQ value function decomposition. Journal of Artificial Intelligence Research,13:227–303

Gabriel S, Zhuang J, Kiet S (2004) A Nash-Cournot model for the north american natural gas market. In Proceeding softhe 6th IAEE European Conference: Modelling in Energy Economics and Policy

Gómez-Sanz J, Fuentes-Fernández R, Pavón J, García Magariño, I (2008) INGENIAS development kit: a visual multi-agent system development environment (BEST ACADEMIC DEMO OF AAMAS' 08). In Proceedings of the Seventh AA-MAS, pages 1675–1676, Estoril, Portugal

Helleboogh A, Vizzari G, Uhrmacher A, Michel F (2007) Modeling dynamic environment sinmulti-agent simulation. JAAMAS, 14(1):87–116

Howard R, Matheson J (1984) Influence diagrams. In Read-ingson the Principles and Applications of Decision Analysis, volume 2, pages 721–762. Strategic Decision Group, Menlo Park, CA

Kitano H, Tadokoro S (2001) RoboCupRescue: A grand challenge for multi-agent systems. Artificial Intelligence Magazine, 22(1):39–52

Malone T, Laubacher R, Dellarocas C (2009) Harnessing crowds: Mapping the genome of collective intelligence. Working paper 2009-001, MIT Center for Collective Intelligence.

Manola, F. And Miller, E. (2004). RDF primer. W3C Recommendation

MIT (2009) MIT Center for Collective Intelligence. http://cci.mit.edu/

OMIP. Iberian Electricity Market Operator. online: 'www.omip.pt'

Pokahr A, Braubach L, Lamersdorf W (2005) A goal deliberation strategy for BDI agent systems. In Proceedings of the Third German Conference on Multi-Agent System Technologies (MATES-2005), pages 82–94. Springer

Prud'hommeaux E, Seaborne A (2008) SPARQL Query language for RDF. W3C Recommendation

Rao A, Georgeff M (1995) BDI agents: From the orytopractice. In Proceedings of the First International Conferenceon Multiagent Systems, pages 312–319, San Francisco, USA

Schuster S, Gilbert N (2004)Simulating online business models. In Proceedings of the 5th Workshop on Agent-Based Simulation (ABS-04), pages 55 – 61

Simari G, Parsons S (2006) On the relationship between MDPs and the BDI architecture. In Proceedings of the Fifth International Joint Conference on Autonomous Agents and Multi agent Systems (AAMAS-06), pages 1041–1048, Hakodate, Japan. ACMPress

Sulis W (1997) Fundamental concepts of collective intelligence. Non linear Dynamics, Psychology, and Life Sciences, (1):35–53

Sutton R, Precup D, Singh S (1999) Between MDPs and semi-MDPs: A frame work for temporal abstraction in reinforcement learning. Artificial Intelligence, 112(1–2): 181–211

Trigo P, Coelho H (2007) Decision making with hybrid models: the case of collective and individual motivations. In Proceedings of the EPIA-07 International Conference (New Trends in Artificial Intelligence), pages 669–680, Guimarães, Portugal

Trigo P, Coelho H (2008a) Decisions with multiple simultaneous goals and uncertaincausal effects. In Artificial Intelligence in Theory and Practice II, volume 276 of IFIP International Federation for Information Processing, pages 13–22. Springer-Verlag.

Trigo P, Coelho H (2008b) Simulating a multi-agent electricity market. In Proceedings of the 1st Brazilian Work shop on Social Simulation (BWSS-08/SBIA-08), Bahia, Brazil

Trigo P, Coelho H (2009) Agent inferencing meets the semantic web. In Progressin Artificial Intelligence, EPIA-09, volume 5816 of Lecture Notes in Artificial Intelligence (LNAI), pages 497–507. Springer-Verlag

Trigo P, Coelho H (2010) Decision making with hybrid models: the case of collective and individual motivations. International Journal of Reasoning-based Intelligent Systems (IJRIS), 2(1):60–72

Trigo P, Jonsson A, Coelho H (2006) Coordination with collective and individual decisions. In Sichman, J.S., Coelho, H., and Rezende, S.O., editors, Advances in Artificial Intelligence, IBERAMIA/SBIA 2006, volume4140 of Lecture Notes in Artificial Intelligence –LNAI, pages 37–47. Springer-Verlag, Ribeirão Preto, Brasil

Trigo P, Marques P, Coelho H (2010) (virtual) agents for running electricity markets. Simulation Modelling Practice and Theory Journal; Simulation-based Design and Evaluation of Multi-Agent Systems, to appear

Venables W, Smith D (2009) An Introduction to R. Network Theory Ltd-publishing free soft ware manuals

Wooldridge M (2000) Reasoning About Rational Agents, chapter Implementing Rational Agents. The MIT Press

Analysis of Crossover Operators for Cluster Geometry Optimization

Francisco B. Pereira[1,2] **and Jorge M. C. Marques**[3]

[1]Instituto Superior de Engenharia de Coimbra, Quinta da Nora, 3030-199 Coimbra, Portugal

[2]Centro de Informática e Sistemas da Universidade de Coimbra (CISUC), 3030-290 Coimbra, Portugal

[3]Departamento de Química, Universidade de Coimbra, 3004-535 Coimbra, Portugal

(e-mail: {xico@dei.uc.pt, qtmarque}@ci.uc.pt)

Abstract We study the effectiveness of different crossover operators in the global optimization of atomic clusters. Hybrid approaches combining a steady-state evolutionary algorithm and a local search procedure are state-of-the-art methods for this problem. In this paper we describe several crossover operators usually adopted for cluster geometry optimization tasks. Results show that operators that are sensitive to the phenotypical properties of the solutions help to enhance the performance of the optimization algorithm. They are able to identify and recombine useful building blocks and, therefore, increase the likelihood of performing a meaningful exploration of the search space.

1 Introduction

An atomic or molecular cluster is composed by a set of particles and it may present distinct physical properties from those of a single molecule or bulk matter. Estimating its most relevant properties has immediate relevance in many areas, ranging from protein structure prediction to the study of the influence of stratospheric clouds in ozone destruction. Also, a proper understanding of cluster properties is crucial for the field of nanotechnology.

The interactions among the particles that compose the cluster may be described by a multidimensional function, designated as Potential Energy Surface (PES), whose knowledge is mandatory in the theoretical study of the properties of a given chemical system [Stillinger 1999]. However, finding an organization for the particles that corresponds to the lowest potential energy turns out to be a NP- hard task [Wille and Vennik 1985]. As this is a critical piece of information, stochastic global optimization algorithms have been increasingly used to discover the global minimum of the PES or, at the least, a set of low energy local minima. For systems with many particles, experiments are usually performed on model functions based on the sum of all pair potentials (i.e., functions that depend only on the dis-

A. Madureira et al. (eds.), *Computational Intelligence for Engineering Systems: Emergent Applications*, Intelligent Systems, Control and Automation: Science and Engineering 46,
DOI 10.1007/978-94-007-0093-2_5, © Springer Science + Business Media B.V. 2011

tance between every pair of particles that compose the aggregate). The Morse function has been widely adopted as it accurately models both long-range interactions, such as those that appear in alkali metal clusters, and short range interactions arising in, e.g., C_{60} molecules [Morse 1929].

Cluster geometry optimization is also relevant for global optimization, as it provides difficult instances for benchmarking new algorithms. Since the early 1990's Evolutionary Algorithms (EAs) have been increasingly applied to this task [Deaven and Ho 1995, Doye et al. 2004, Grosso et al. 2007, Hartke 2001, Johnston 2003, Pereira and Marques 2009, Zeiri 1995]. State-of-the-art methods usually combine a global evolutionary component with a gradient-driven local search procedure. A real valued representation and specific genetic operators help to enhance the effectiveness of the algorithm. In what concerns crossover, the Cut and Splice operator (C&S) was proposed in 1995 by [Deaven and Ho 1995] and, since then, it has been adopted by most of the evolutionary approaches for cluster geometry optimization. C&S is sensitive to the semantic properties of a solution and therefore is able to perform a meaningful recombination of the parents. Recently we proposed Generalized Cut and Splice crossover (GenC&S), an extension that attempts to remove some limitations that are linked to the application of C&S [Pereira et al. 2008]. An empirical locality analysis was performed with both operators and confirmed that GenC&S exhibits a behavior that is more in accordance to what is expected from crossover. GenC&S was recently applied to argon clusters and it was able to discover all putative global optima up to 78 atoms [Marques et al. 2008]. Additionally, a variant of GenC&S for heterogeneous clusters is proposed in [Marques and Pereira 2010].

In this paper we apply both C&S and GenC&S operators to several difficult instances of short-ranged Morse clusters to gain a deeper insight of their optimization abilities. A detailed analysis of the results will help to understand why GenC&S is more effective for cluster geometry optimization.

The structure of the paper is the following: in section 2 we present the potential function used to model Morse clusters. Section 3 comprises the presentation of the optimization algorithm used in the experiments, including a detailed description of the crossover operators. The main results are presented and discussed in section 4. Finally, section 5 gathers the main conclusions.

2 Morse Potential

The energy of Morse clusters is obtained by summing the Morse potentials describing all pair-wise interactions that occur in the aggregate [Morse 1929]:

$$V_{Morse} = \varepsilon \times \sum_{i}^{N-1} \sum_{j>i}^{N} \left(e^{-2\beta(r_{ij}-r_0)} - 2e^{-\beta(r_{ij}-r_0)} \right) \quad (1)$$

In equation (1), r_{ij} stands for the Euclidean distance between atoms i and j, ε is the bond dissociation energy, r_0 is the equilibrium bond length and β is the range exponent of the potential. We adopt the scaled version of the Morse potential with non-atom specific interactions and, therefore, both ε and r_0 are set to 1 [Doye and Wales 1997]. The only adjustable parameter is then β, which determines the range of interparticle forces. We set β=14, corresponding to a short-range interaction. Discovering the global minimum is particularly hard these instances, as the search landscape is extremely rough and it contains a huge number of local minima [Doye et al. 2004]. Furthermore, the global minima of instances with a different number of atoms correspond to structures with distinct structural organization and properties, preventing the application of biased methods [Doye and Wales 1997, Grosso et al. 2007].

2.1 Related Work

An historical description of the application of EAs to cluster geometry optimization problems can be found in [Hartke 2004]. In this section we briefly highlight the most relevant contributions, focusing our attention on Morse clusters.

Doye and Wales were the first ones to apply an optimization algorithm to Morse clusters. They developed a basin-hopping method (it combines a Monte Carlo algorithm with local search) and were able to discover all but 12 of the putative global optima for short-ranged Morse clusters up to 80 atoms [Doye and Wales 1997]. Johnston proposed the first hybrid EA for this problem. It was applied to short and medium range Morse clusters and discovered all putative global optima until 50 atoms [Johnston 2003]. In 2002, Locatelli and Schoen proposed population-basin hopping with a two-phase local optimization [Locatelli and Schoen 2002]. This hybrid method proved to be extremely effective in discovering all putative global optima of Morse clusters until 80 atoms [Grosso et al. 2007]. The shortcoming is this approach is that the first phase of local search requires the specification of a set of parameters that are problem dependent and therefore it cannot be easily extended to situations where the global optimum is unknown. Two recent papers describe approaches to remove the bias from the first phase of local search [Cassioli et al. 2008, Pereira and Marques 2008]. In 2009, Pereira and Marques proposed a simple hybrid EA that is able to discover all putative global optima for short-ranged Morse clusters until 80 atoms [Pereira and Marques 2009]. This is the first completely unbiased population-based method that was able to succeed in this task. Its effectiveness is related to the adoption of appropriate distance measures that maintain the diversity in the population.

3 Hybrid Optimization Algorithm

The optimization framework adopted in this work combines a steady-state EA and a local search procedure. It is well-known from previous studies that hybrid approaches are particularly effective in discovering high quality solutions for cluster geometry optimization problems [Deaven and Ho 1995, Doye et al. 2004, Hartke 2001, Johnston 2003]. On the one hand, the EA is able to efficiently sample large areas of the search space. On the other hand, a local search method that relies on first order derivative information, effectively guides search into the nearest local optimum. We rely on the Broyden-Fletcher- Goldfarb-Shanno method (L-BFGS), a quasi-Newton conjugate gradient procedure [Liu and Nocedal 1989]. L-BFGS is applied to every new solution generated by the EA.

3.1 Evolutionary Algorithm

A straightforward steady-state model is used. Unlike generational EAs, in this model populations overlap and descendants compete for survival with existing solutions. In each iteration, a sequence of steps are performed: i) tournament selection chooses two parents that will be used to create the offspring; ii) crossover and mutation are applied to the selected individuals (see next section for details about the genetic operators); iii) Offspring are locally optimized and evaluated; iv) replacement rules are applied to determine if the offspring will be allowed to join the population.

There are two main driving forces that regulate the replacement of solutions: quality and diversity. A distance measure must be defined to estimate the similarity between two solutions. In this work we rely on the center of mass distance, a measure that is based on the distance of atoms to the cluster's center of mass [Grosso et al. 2007]. Results reported in [Grosso et al. 2007; Pereira and Marques 2009] confirm that it is effective in accessing the similarity of possible solutions for Morse instances. In general terms, a descendant D is allowed to enter the population if one of the following conditions is satisfied (consult [Lee et al. 2003] for a detailed description of the replacement strategy):

i) D is similar to a solution X that belongs to the current population and D is better than X. If this happens, D replaces X;

ii) D is different from all solutions currently in the population. In this case, it replaces the worst solution X, providing that D is better than X.

The application of this replacement strategy requires the definition of a parameter, d_{min}, specifying the minimum allowed distance between any two solutions that simultaneously belong to the population.

3.1.1 Representation and Genetic Operators

Each possible solution encodes the location of the particles that compose the cluster. For an instance with N atoms, it encodes 3×N real values specifying the Cartesian coordinates of each particle. The domain for all variables is [0, $N^{1/3}$] [Johnston 2003]. Only inter-atomic distances larger than 0.5 are allowed, since the potential becomes too repulsive is two atoms approach too much. Fig. 5.1 exemplifies the structure of the chromosome.

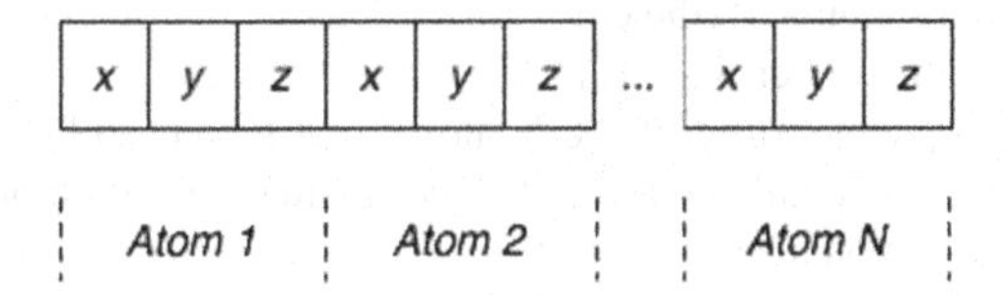

Fig. 5.1 Structure of a chromosome

Both crossover and mutation are applied to generate descendants. The main goal of this paper is to study the effect of different crossover operators in the results achieved by the optimization algorithm. C&S is an operator that is sensitive to the semantic properties of individuals involved in a crossover operation [Deaven and Ho 1995].It selects sub-clusters of the parents and combines them in such a way that it increases the likelihood of preserving relevant features of promising solutions. Since its proposal, C&S has been adopted by many researchers that applied EAs to cluster optimization problems [Hartke 2001; Johnston 2003]. Results confirm that it contributes to enhance the efficacy of the optimization algorithm. In detail, C&S performs the following operations to generate descendant D1 from parents P1 and P2 (another descendant D2 is created just by swapping the roles played by P1 and P2):

1. Apply random rotations to P1 and P2;
2. Define a random horizontal cutting plane (parallel to the *xy* plane) for P1. This plane splits P1 in two complementary parts (X atoms below the plane and N-X atoms above it);
3. Define a horizontal cutting plane (parallel to the *xy* plane) for P2, in such a way that X atoms stay below the plane and N-X are above it;
4. Generate D1 and D2 by combining complementary parts of each one of the parents.

Special precautions are taken when merging sections from different parents to ensure that the distance between two atoms in D1 is never below 0.5. C&S ensures that the contribution of each parent is composed by a subset of atoms that are close together (i.e., a sub-cluster) and, therefore, they will tend to have a low potential energy. Hopefully, these building blocks increase the likelihood of discovering promising solutions for the problems being solved.

Recently we proposed GenC&S, an extension of the original C&S operator [Pereira et al. 2008]. The most important difference between these two variants is that GenC&S relaxes some constraints that exist in the original version of the operator (e.g., the requirement of defining a horizontal plane to determine which atoms are inherited from each one of the parents). It just considers distances between atoms to determine the subset of atoms from each parent that will be used to create a descendant. Additionally, GenC&S does not require the application of random rotations to the parents involved in a crossover operation. Considering the optimization of a cluster with N atoms, parents P1 and P2 lead to the creation of descendant D1 after the following steps are performed:

1. Select a random atom CP (the cut point) in P1.
2. Select a random number $S \in [1, N-1]$.
3. Create a list L_{P1} consisting of the N atoms that belong to P1. Items in L_{P1} are ordered according to an increasing distance from the location of CP.
4. Copy the first S atoms from L_{P1} to D1.
5. Create a list L_{P2} consisting of the N atoms that belong to P2. Items in L_{P2} are ordered according to an increasing distance from the location of CP.
6. Remove from L_{P2} atoms that are too close (i.e., at a distance smaller than 0.5) to particles already copied to D1.
7. Copy the first (N-S) atoms from L_{P2} to D1.
8. If less than N atoms were copied to D1 then the individual is completed with particles placed at random locations.

The example from Fig. 5.2 illustrates a crossover situation between two individuals with 5 atoms. Although it is presented in a two-dimensional plane, it clearly describes the crossover operation that occurs in three dimensions. In panel a) the two parents are displayed. In this example CP is atom number 5 from P1 and S=3. The dotted circle in P2 marks the location of atom 5 from P1 (the cut point). This position is needed later to determine which atoms from P2 will be passed to the descendant. In panel b) we present the atoms inherited by D1 from P1. The 3 closest atoms to the cut point were selected. In panel c) we display the overlapping between the atoms that were already copied to D1 and those that belong to P2. We consider that overlapping atoms are too close and therefore cannot belong to the same cluster. In this example, atoms 8 and 9 from P2 are too close to atoms that already belong to D1 (so they will not be inherited by the descendant, even though they are near the cup point). In panel d) we show the atoms from P2 that can be inherited by the descendant. Finally, in panel e) we present the final descendant after receiving the 2 remaining atoms from P2. This parent passed atoms 6 and 7 to D1, since they are the closest non-overlapping atoms to the cut point.

For completeness, in the next section we also present results obtained with standard uniform crossover. With this operator, parents are scanned left to right and, in each position, the descendant inherits the atom from one of the parents with equal probability. The operator ensures that the child does not receive atoms that are too close to each other. Uniform crossover is blind to the spatial distribution of atoms. By comparing its results with those achieved by C&S and GenC&S we will gain insight into the relevance of developing crossover operators that are sensitive to the structure of the solutions.

Sigma mutation is adopted to modify the location of a particle inside the cluster. The new position is obtained by slightly perturbing each coordinate with a random value drawn from a Gaussian distribution with mean 0 and standard deviation σ (σ is a parameter of the algorithm).

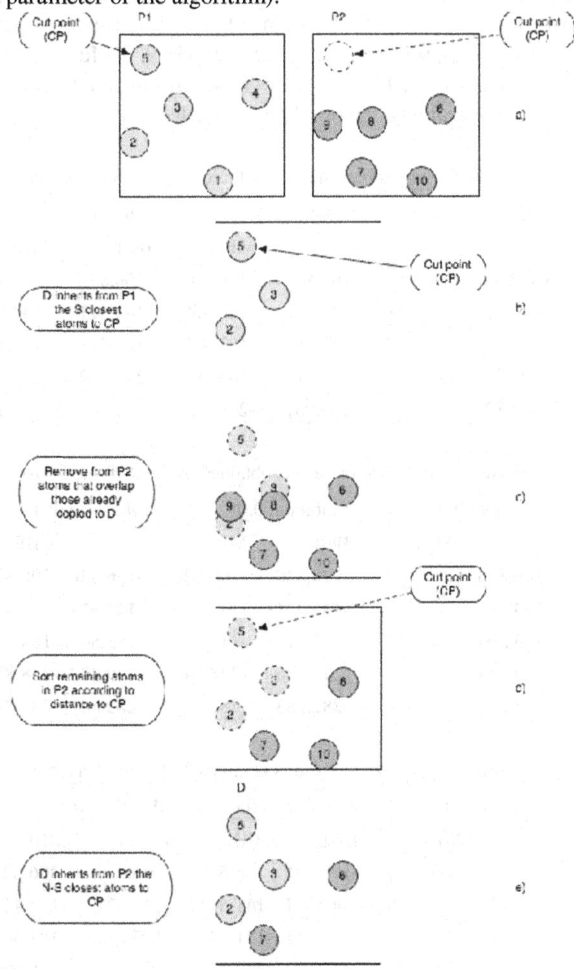

Fig. 5.2 Application of Generalized Cut and Splice crossover between two clusters. In this example N=5 and Cut size=3.

4 Results and Discussion

The performance of the hybrid EA with different crossover operators is accessed on a selected subset of short-ranged Morse instances. More specifically we selected the instances with the following number of atoms N= {30, 38, 47, 61, 68}. This set includes examples that are considered to be difficult to optimize (see, e.g., [Grosso et al. 2007, Pereira and Marques 2009]).

The settings of the algorithm are the following: Number of runs: 30; Population size: 100; Evaluations: 5,000,000; Tournament size: 5; Crossover operators: {Uniform, C&S, GenC&S}; Crossover rate: 0.7; σ: $0.05 \times N^{1/3}$; Mutation rate: {0, 0.05, 0.1}. Each iteration performed by the L-BFGS algorithm counts as one evaluation. Initial populations were always randomly generated.

Table 5.1 Optimization results obtained by C&S crossover.

	Mutation: 0.0		Mutation: 0.05		Mutation: 0.1	
N	Best	MBF	Best	MBF	Best	MBF
30	**-106.836**	-106.756	**-106.836**	-106.704	**-106.836**	-106.736
38	**-144.321**	-142.568	**-144.321**	-142.439	**-144.321**	-142.617
47	-183.411	-182.442	-183.411	-182.452	-183.411	-182.546
61	-249.586	-246.093	-248.612	-246.209	-248.612	-247.157
68	-282.682	-279.135	-280.961	-279.201	-282.682	-279.105

Table 5.2 Optimization results obtained by GenC&S crossover.

	Mutation: 0.0		Mutation: 0.05		Mutation: 0.1	
N	Best	MBF	Best	MBF	Best	MBF
30	**-106.836**	-106.833	**-106.836**	-106.828	**-106.836**	-106.821
38	**-144.321**	-143.999	**-144.321**	-143.366	**-144.321**	-143.428
47	**-183.508**	-183.098	-183.411	-183.156	**-183.508**	-183.191
61	**-249.588**	-248.745	-249.586	-248.741	**-249.588**	-248.833
68	**-282.683**	-282.416	**-282.683**	-282.357	**-282.683**	-282.211

Table 5.3 Optimization results obtained by Uniform crossover.

	Mutation: 0.0		Mutation: 0.05		Mutation: 0.1	
N	Best	MBF	Best	MBF	Best	MBF
30	-106.836	-106.522	-106.836	-106.540	-106.836	-106.621
38	-144.321	-141.633	-144.321	-141.512	-144.321	-141.414
47	-183.399	-181.057	-183.411	-181.042	-183.411	-181.342
61	-249.153	-245.402	-247.593	-245.773	-249.153	-245.920
68	-282.624	-277.172	-282.682	-278.436	-282.604	-278.389

In tables 5.1, 5.2 and 5.3 we present the results achieved by the three versions of the hybrid EA (one for each crossover operator). Columns *Best* show the potential energy of the best solution found, whilst columns *MFB* present the mean best fitness (best solutions found averaged over 30 runs). In each table we present results achieved by the three different mutation rates. Entries in bold highlight experiments that were able to discover the putative optimum.

Results clearly show that both C&S and GenC&S outperform uniform crossover, confirming that operators sensitive to the phenotypic properties of a solution lead to an enhanced performance of the optimization algorithm. In what concerns the two specific operators for cluster geometry optimization, GenC&S obtains better results showing that the removal of some artificial constraints helps the algorithm to effectively explore the search space. For the three largest instances, only GenC&S allowed the hybrid EA to discover the global optimum.

Another interesting finding is that the hybrid EA is not sensitive to the variation of mutation rate. In fact, results even show that mutation is not needed at all, as the outcomes obtained without mutation are similar to those achieved when this operator is included in the EA. The explanation for this result is probably related to the application of local search that might be seen as some kind of directed mutation. When L-BFGS is applied it slightly modifies the location of the atoms (trying to find a more suitable configuration) and, therefore, its effect is similar to that of sigma mutation. Also, the EA includes a diversity mechanism, which naturally prevents the population from premature convergence.

In the charts from Fig. 5.3 we present the evolution of the MBF in tests performed with the Morse instance with 47 atoms (results for other instances follow the same trend): panel a) displays results obtained with C&S operator, panel b) concerns GenC&S and panel c) shows the progress of MBF when uniform crossover is used. They support the previous analysis, as the evolution of the MBF is similar for the three mutation rates adopted. We performed a simple statistical analysis that confirms that there are never significant differences in the results achieved by the same crossover operator when different mutation rates are used (t-test with level of significance 0.01).

The chart from Fig. 5.4 displays the evolution of the MBF when different crossover operators are used. The Morse instance with 47 atoms and a mutation rate of 0.05 is considered, although the results are similar to those that occur with other instances and/or mutation rates. The performance ranking is clearly visible in this figure. The hybrid EA with uniform crossover exhibits that worst performance, whilst GEnC&S is the operator that helps the optimization algorithm to achieve better results. C&S crossover is between the two extremes. Again, we performed a simple statistical analysis that, for the most cases, confirms the existence of significant differences in the results achieved by different crossover operators when applied to the same instance (t-test with level of significance 0.01).

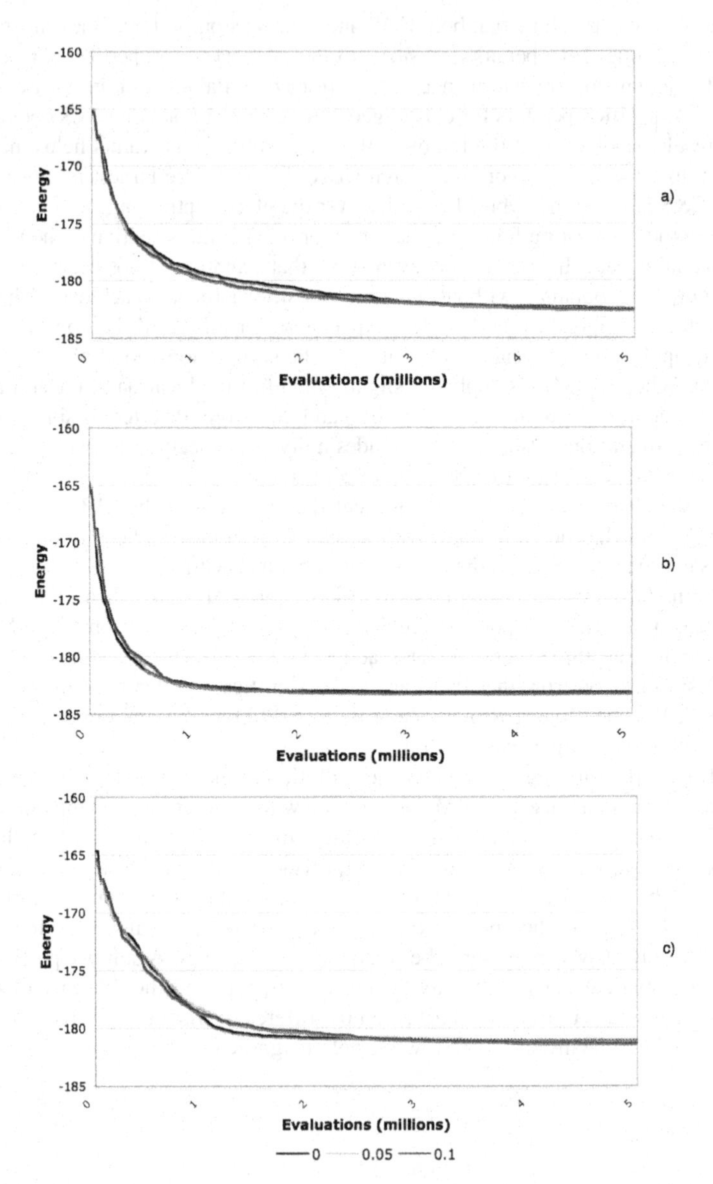

Fig. 5.3 Evolution of MBF in the optimization of the Morse cluster with 47 atoms. Three panels are presented: a) displays results obtained with C&S crossover, b) with GenC&S crossover and c) with Uniform crossover.

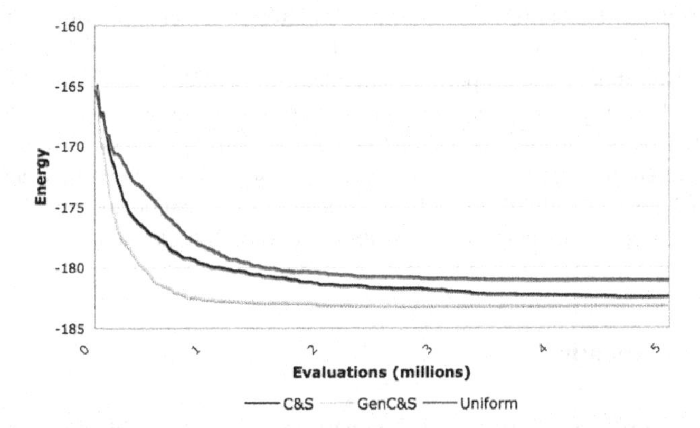

Fig. 5.4 Evolution of MBF in the optimization of the Morse cluster with 47 atoms in experiments performed with three different crossover operators. The mutation rate is 0.05.

In table 4 we present a final set of results. Here we show (again for the instance with 47 atoms), for all combinations of crossover operators and mutation rates, the contribution of genetic operators to the discovery of new best solutions. Column Total displays how many new best solutions, (i.e., solutions that are better than any other discovered in that run) were found by the EA during the optimization. Results are totals for the 30 runs. Columns Mutation and Crossover show the proportion of new best solutions that were discovered thanks to, respectively, mutation and crossover (e.g., when C&S crossover and a mutation rate of 0.05 are adopted, sigma mutation contributes to discover 91% of the new best solutions and C&S helps to create 80%).

Table 5.4 Contribution of the genetic operators for the discovery on new best solutions in the optimization of the Morse cluster with 47 atoms.

	M_Rate	Total	Mutation	Crossover
	0.0	538	0.00	1.00
C&S	0.05	504		0.91 0.80
	0.1	494	1.00	0.72
	0.0	445	0.00	1.00
GenC&S	0.05	436	0.92	0.96
	0.1	440	1.00	0.91
	0.0	445	0.00	1.00
Uniform	0.05	499	0.93	0.85
	0.1	517	1.00	0.83

A brief perusal of the table reveals that the outcomes are similar for the three crossover operators. As expected, the increase in the mutation rate leads to a more frequent contribution of sigma mutation to the discovery of new best solutions.

Crossover exhibits a high degree of participation in the creation of new best solutions. This is true for all operators (with a single exception, it is always above 80%). Even though the proportions obtained for GenC&S are slightly higher than those of both C&S and uniform crossover, results suggest that the difference in performance is probably not related to the active participation of these operators in the exploration of the search space. The increased effectiveness of GenC&S is a consequence of its ability to perform a meaningful recombination of the building blocks that compose the promising solutions chosen by selection.

5 Conclusions

Hybrid algorithms combining evolutionary methods and local search procedures are state-of-the-art techniques for cluster geometry optimization. A key component of the EA is the crossover operator that identifies and combines useful building blocks that may help to discover high quality solutions. In this paper we introduced three crossover operators that can be used in this problem and described a set of experimental results that helped to gain insight into the weaknesses and strengths of each one of the operators.

Results clearly show that operators that are sensitive to the semantic properties of the individuals they manipulate are better suited for this optimization task. Of all operators tested, GenC&S crossover proved to be the most effective. This operator, not only is able to see the individual as a 3D cluster, but also removes some artificial constraints that existed in previous operators.

Acknowledgments This work was supported by Fundação para a Ciência e Tecnologia, Portugal, under grant PTDC/QUI/69422/2006. We are grateful to the John von Neumann Institut für Computing, Jülich, for the provision of supercomputer time on the IBM Regatta p690+ (Project EPG01).

References

Cassioli A, Locatelli M, Schoen F (2008) Global optimization of binary Lennard-Jones clusters. Optimization Methods and Software, online December 2008 (DOI: 10.1080/10556780802614101)

Deaven D, Ho K (1995) Molecular geometry optimization with a genetic algorithm. Phys Rev Lett, 75:288–291

Doye JPK, Leary M, Locatelli M, Schoen F (2004) Global optimization of Morse clusters by potential energy transformations. Informs Journal on Computing, 16:371–379

Doye JPK, Wales D (1997) Structural consequences of the range of the inter-atomic potential. a menagerie of clusters. J Chem Soc Faraday Trans, 93:4233–4243

Grosso A, Locatelli M, Schoen F (2007) A population-based approach for hard global optimization problems based on dissimilarity measures. Math Program Ser A, 110:373–404

Hartke B (2001) Global geometry optimization of atomic and molecular clusters by genetic algorithms. In Proceedings of the Genetic and Evolutionary Computation Conference (GECCO-2001), 1284–1291. Morgan-Kaufmann

Hartke B (2004) Application of evolutionary algorithms to global cluster geometry optimization. In Applications of Evolutionary Computation in Chemistry, Structure and Bonding, 33–53. Springer-Verlag

Johnston R (2003) Evolving better nanoparticles: Genetic algorithms for optimising cluster geometries. Dalton Transactions, 22:4193–4207

Lee J, Lee IH, Lee J (2003) Unbiased global optimization of Lennard-Jones clusters for $n \leq 201$ by conformational space annealing method. Phys Rev Lett, 91:080201.1–080201.4

Liu DC, Nocedal J (1989) On the limited memory method for large scale optimization. Mathematical Programming B, 45:503–528

Locatelli M, Schoen F (2002) Fast global optimization of difficult Lennard-Jones clusters. Comput Optim Appl, 21:55-70

Marques JMC, Pereira FB (2010) An Evolutionary Algorithm for Global Minimum Search of Binary Atomic Clusters. Chem Phys Lett, 485:211-216

Marques JMC, Pereira FB, Leitão T (2008) On the use of different potential energy functions in rare-gas cluster optimization by genetic algorithms: application to argon clusters. Journal of Physical Chemistry A, 112:6079-6089

Morse P (1929) Diatomic molecules according to the wave mechanics. ii. vibrational levels. Phys Rev, 34:57–64

Pereira FB, Marques JMC (2008) A self-adaptive evolutionary algorithm for cluster geometry optimization. In Proceedings of the Eight International Conference on Hybrid Intelligent Systems, 678–683. IEEE Press

Pereira FB, Marques JMC (2009) A study on diversity for cluster geometry optimization. Evolutionary Intelligence, 2:121-140

Pereira FB, Marques JMC, Leitão T, Tavares J (2008) Efficient evolutionary algorithms for cluster optimization: A study on locality. In Advances in Meta-heuristics for Hard Optimization, 223–250. Springer-Verlag

Stillinger F (1999) Exponential multiplicity of inherent structures. Phys Rev E, 59:48-51

Wille LT, Vennik J (1985) Computational complexity of the ground-state determination of atomic clusters. J Phys A: Math Gen 18:419-422

Zeiri Y (1995) Prediction of the lowest energy structure of clusters using a genetic algorithm. Phys Rev, 51:2769–2772

A Support Vector Machine based Framework for Protein Membership Prediction

Lionel Morgado[1], Carlos Pereira[2], Paula Veríssimo[3] and António Dourado[4]

[1,4]Center for Informatics and Systems of the University of Coimbra

Polo II – University of Coimbra, Portugal

(email: {lionel;cpereira;dourado}@dei.uc.pt)

[2]Coimbra Institute of Engineering - ISEC

Quinta da Nora, 3030-199 Coimbra, Portugal

(email: cpereira@isec.pt)

[3]Department of Biochemistry and Center for Neuroscience and Cell Biology

University of Coimbra, 3004-517 Coimbra, Portugal

(email: paulav@ibili.uc.pt)

Abstract The support vector machine (SVM) is a key algorithm for learning from biological data and in tasks such as protein membership prediction. Predicting structural information for a protein from its sequence alone is possible, but the extreme data complexity demands kernels with a dedicated design like the state-of-the-art profile kernel that exploits a very large feature space. Such a huge representation and the enormous data bases used in proteomics require an effort mirrored in an increased processing time that must be reduced to an acceptable amount. Considering the present computation paradigm, the implementation of such systems shall take advantage of parallelization and concurrency. A special machine learning architecture based on SVM binary models and a neural network (NN) is proposed to handle the large multiclass problem of protein superfamily prediction, and parallelized through a multi-agent strategy that uses JADE (Java Agent DEvelopment Framework) to reduce the total processing time when getting a prediction for a new query protein. The efficiency of the algorithm and the advantages of the parallelization are shown.

Keywords Protein, SVM, NN, Parallelization, JADE

A. Madureira et al. (eds.), *Computational Intelligence for Engineering Systems: Emergent Applications*, Intelligent Systems, Control and Automation: Science and Engineering 46,
DOI 10.1007/978-94-007-0093-2_6,

1 Introduction

A traditional issue in bioinformatics is the classification of protein sequences into functional and structural groups based on sequence similarity. Approaches for protein classification through homology can be divided into three main groups: pairwise sequence comparison methods, generative models and discriminative classifiers. The most successful methods for homology detection are the discriminative, that combine SVMs [Vapnik 1998] with special kernels [Jaakkoola et al. 2000, Leslie et al. 2002, Weston et al. 2003, Kuang et al. 2005]. The SVM is a powerful machine learning technique that unites high accuracy with good generalization, achieving state-of-the-art results. However, the SVM implementations present some processing limitations when faced with data projected to very high dimensional feature spaces and dealing with the numerous examples and large multi-class problems presented in biology. Such harsh conditions rise the total processing time needed to train the discriminative models and when calculating a prediction for a new protein which membership is unknown. This may be the main reason why large SVM based systems are not very common and available. Considering the current computational paradigm, distributing the calculations among several machines or processing units can bring computation benefits. Obviously, breaking a problem into subproblems and solving them independently requires synchronization between the entities whose results must be combined to achieve the final solution. One way to strike all these issues can be through the use of artificial intelligence techniques such as a multi-agent society. Programmatically an agent can be defined as an application that runs independently and executes one or several tasks according to its more or less complex knowledge of the world, embedding the ability to communicate and interact with other agents or entities. An advantage from following this approach is the availability of free software development platforms such as JADE (http://jade.tilab.com) to launch and manage societies of agents, either with local and remote features which allows the developer to focus on the system itself instead of the programmatic details of the messaging and synchronization. This is a low cost solution that does not demand the acquisition of special hardware, and rationalizes the computational resources available that many times are not completely used. The scalability of the system is a plus.

These factors took us to choose this strategy to develop the prototype application for protein multi-class classification reported here. The framework uses a benchmark data set from SCOP [Murzin et al. 1995] previously used on remote homology detection. Its architecture is exposed and its performance is studied showing its ability to discriminate, and the advantages of the parallelized computation.

An overview of SVMs and the profile kernel for protein classification is presented in Section 2, accompanied by a reference to the kernel trick which is fundamental for the decomposition of the computation. Section 3 is dedicated to the description of the system design, namely the multi-class discriminative algorithm developed and the multi-agent architecture. Section 4 presents the experiments concerning the accuracy and speedup, as well the respective analysis. Final conclusions and references to future work are given in the last Section.

2 SVMs with profile kernel

The SVM belongs to a successful class of learning systems strictly connected with statistical learning theory. It tries to find a function f that minimizes the expected risk:

$$R[f] = \iint L(y, f(x))dP(y \mid x)dP(x) \tag{1}$$

where the distribution of the examples $P(x)$ and their classifications $P(y \mid x)$ are unknown, and where L is a loss function that measures the error of predicting y with $f(x)$.

Based on this principle the SVM builds a classifier that generally can be described by:

$$f(x) = w.\phi(x) + b \tag{2}$$

for a training data set $\{(x_i, y_i) \in \Re^n \times \{-1,1\} \forall i \in \{1,\dots,N\}\}$ from a binary problem.

A transformation is performed in order to change an eventually non-linear input problem representation to a linear one in a given feature space ϕ. This mapping is implicit in a kernel function $K(x_i, x_j)$ given by the dot product:

$$K(x_i, x_j) = \phi(x_i).\phi(x_j) \tag{3}$$

Considering the distance of a point to the decision hyperplane, it is possible to define a margin with width $\frac{2}{\|w\|^2}$, from which the optimisation process comes:

$$\min_{w,b,\xi} \frac{1}{2}\|w\|^2 + C\sum_{i=1}^{N} \xi_i^p \tag{4}$$

, with typical order values $p = \{1, 2\}$, and subject to:

$$y_i(w\phi(x_i)+b) \geq 1-\xi_i, \tag{5}$$

and

$$\xi_i \geq 0, \forall i \in \{1,...,N\} \tag{6}$$

where $C>0$ is a trade-off parameter that controls the relative importance of minimizing the norm of w, and ξ a variable that minimized the upper bound of the empirical risk in the called soft-margin formulation. The slack ξ gives more flexibility to a model by adding tolerance to outliers and noise in the training data set.

The mapping function isn't known in the optimisation process, so it can't be made directly for variables w and b. In the traditional SVM the Lagrangian plays an important role in this phase. This problem can then be expressed as a quadratic program in the dual form:

$$\min_{0\leq\alpha_i\leq C} W = \frac{1}{2}\sum_{i,j=1}^{N} \alpha_i Q_{ij} \alpha_j - \sum_{i=1}^{N} \alpha_i + b\sum_{i=1}^{N} y_i \alpha_i \tag{7}$$

where the α parameters are Lagrange coefficients and $Q_{ij} = y_i y_j \phi(x_i)\phi(x_j)$, with $\forall i \in \{1,\ldots,N\}$ and $\forall j \in \{N\}$.

The collection of points for which $\alpha \neq 0$ describe the decision frontier and each particular instance is termed a support vector.

Employing inner products in feature space and taking advantage of the kernel function to compute them is called the kernel trick. Using the kernel trick it is possible to simplify the computation, particularly in extremely high-dimensional feature spaces typical for example in string kernels. Nevertheless, all calculations in the feature space must be devised in terms of inner products.

Several kernels have been proposed for protein classification using sequence alone [Jaakkoola et al. 2000; Leslie et al. 2002; Weston et al. 2003; Kuang et al. 2005]. The kernel function aims emphasizing important biological information while converting variable length strings that represent proteins, into numeric fixed size feature vectors. This mapping is mandatory in the sense that the learning machine demands feature vectors with a fixed number of attributes. It also has a strong impact in the final accuracy and the complexity of the learned models.

The profile kernel [Kuang et al. 2005] is a state-of-the-art string kernel used in proteomics that acts over an input space composed of all finite sequences of characters from an alphabet A with l elements, changing it to a feature space with l^k dimensions that represent all the possible k-length contiguous subsequences that may be contained in a protein.

The first stage to produce a profile kernel consists in determining the profiles *P(x)* of a sequence *x*. Profiles are statistically estimated from close homologues stored in a large sequence database, and can be defined as:

$$P(x) = \{p_i(a), a \in A\}_{i=1}^{N} \tag{8}$$

with p_i the emission probability of amino acid a in position i and $\sum_{a \in A} p_i(a) = 1$ for every position. These profiles are related with the probability of a given amino acid in a given position of the query sequence to be replaced by another one in the same position, therefore expressing the tendency for a mutation to occur. Considering the profile for a particular protein, it is possible to account all subsequences that can be present not only due to mere change, and for which the score is over a given threshold σ.

The profile feature mapping is defined as:

$$\Phi_{(k,\sigma)}^{profile}(P(x)) = \sum_{j=0\ldots|x|-k} \left(\phi_\beta(P(x[j+1:j+k]))\right)_{\beta \in A^k} \tag{9}$$

where the coordinate $\phi_\beta(P(x[j+1:j+k])) = 1$ if β belongs to the allowed subsequences. Otherwise, the coordinate assumes the value 0.

Finally, the profile kernel is given by the dot product:

$$K_{(k,\sigma)}^{profile}(P(x), P(y)) = \left\langle \Phi_{(k,\sigma)}^{profile}(P(x)), \Phi_{(k,\sigma)}^{profile}(P(y)) \right\rangle. \tag{10}$$

3 System architecture

The system developed allows getting a membership prediction for an unknown sequence from its sequence alone. For that it uses a SVM based multi-class discriminative frame with the state-of-the-art profile kernel. The algorithm is divisible in several steps that can be performed independently. To parallelize and coordinate the prediction procedure a multi-agent system was developed using the JADE platform. The objective of such strategy is to reduce the amount of processing time, as later described in Section 4.

3.1 The protein membership prediction algorithm

The SVM formulation for multiclass training is a reality [Crammer and Singer, 2001; Joachims et al. 2009], but the extra processing effort that it brings for very large multiclass problems compared to architectures that combine binary models for the same purpose, makes the former pretermitted to the latter. Literature [Rifkin et al. 2003; Hsu and Lin 2002] shows two main methods for multi-class classification using binary models: one-versus-the-rest and all-versus-all. In the one-versus-the-rest architecture N one-versus-the-rest classifiers are trained, being N the total number of classes to be discriminated and where each model is built to distinguish between the instances of interest and all the other groups. On the other hand, the all-versus-all design assembles $N(N-1)/2$ binary models, each of which aims to distinguish between two different membership groups. Both strategies can use similar techniques to combine the outputs of the individual binary classifiers to get the final prediction. Occasionally, a voting system is a good option to get the verdict according to the class that gathers the major agreement. However, the outputs from the binary models are frequently not comparable. Because of this adversity it is commonly used another classifier in the top of the ensemble to establish a relationship between the outputs of the binary classifiers and eventually to execute some learning corrections.

The architecture of the discriminative algorithm developed follows a one-versus-the-rest philosophy and it is described in Fig. 6.1. During an initial phase the query string sequence is subjected to PSI-BLAST [Altschul et al. 1997], to get the profile needed to build the profile kernel feature space. The huge dimentionality of the feature space creates a bottleneck in the dot product computation. To decompose the algorithm and eliminate redundant calculations brought by the one-versus-the-rest strategy selected, the kernel trick principles are invoked, so naturally, the following step consistes in determining the dot product between all the instances to be included in each SVM kernel that will be used in the training phase.

Unfortunately, because the training data set partition is extremely unbalanced and the presence of overlapping classes induces the learning algorithm to define biased decision hyperplanes to promote global accuracy in detriment of the correct identification of the less numerous positive instances. For that reason, a very high false negative rate can be observed when using the binary classifiers. To solve this problem and also to establish a correct relationship between the outputs of the SVM models a feed-forward neural network with only one layer is used on top of them.

Learning from the outputs is an idea presented in the error-correcting output codes [Dietterich, and Bakiri 1995; Crammer and Singer 2000]. However, instead of learning from the predicted labels we use richer information: the decision values. A decision value grants to the neural network information concerning the distance of a particular instance to the decision hyperplane rather than a simplifyed indication about the side of the frontier in which it lays. In the end, a final prediction for the protein superfamily is obtained. Associated to the superfamily are upper hierarchies such as the fold and class to which information can also be retrieved.

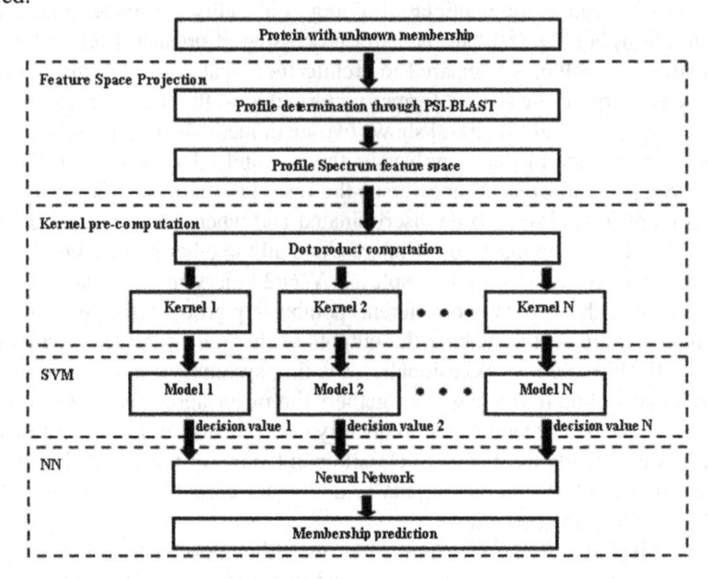

Fig. 6.1 Structure of the prediction algorithm

3.2 Multi-agent implementation

During this work was developed a multi-agent application that allows parallelizing and distributing the computation among different machines. For that objective was used the JADE framework. JADE provides a runtime environment that once activated in the host machine allows creating agents in containers that can be distributed between several processing units, also enabling their communication using FIPA (Foundation for Intelligent Physical Agents) Agent Communication Language standards through a peer-to-peer network architecture. Each agent has a queue messaging system supported by the JADE runtime, however when, how and which message must be sent is defined programmatically. Whenever a message is posted in the message queue the receiving agent is notified. JADE has native agents that aim to facilitate the management of the system, allowing the developer to focus on the framework design instead of the communication details. The Agent Management System is mandatory and responsible for the supervision of the access and use of the agent platform. Every agent must be registered with an Agent Management System in order to get a valid agent identifier. The Directory Facilitator provides a yellow pages service, by which it is possible to find a specific kind of agent according to an attribute associated during registration. Finally, the Agent Communication Chanel controls the exchange of messages within the platform, whether they are or not to/from remote platforms.

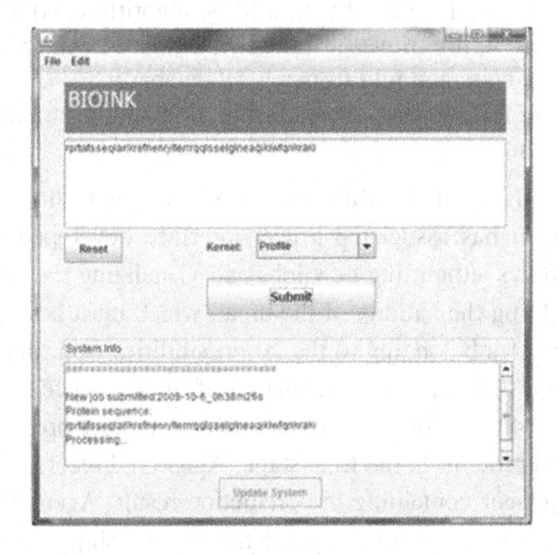

Fig. 6.2 The user interface coupled to AgentGUI

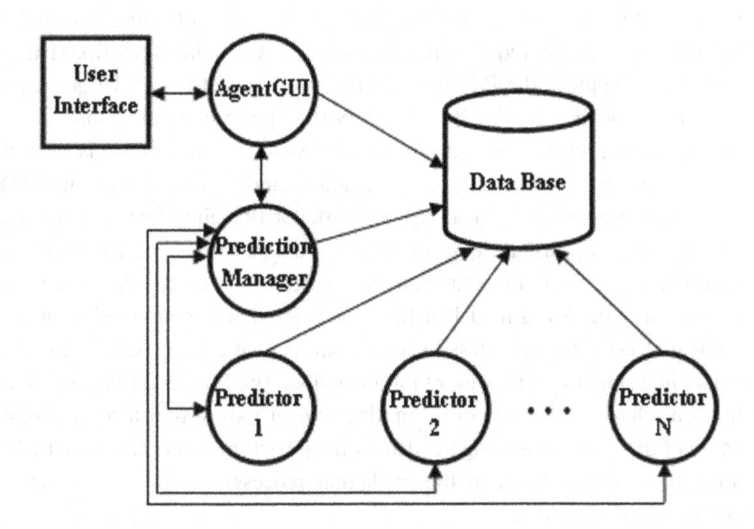

Fig. 6.3 Architecture of the multi-agent implementation. Agents are represented by circles.

The application that was created allows the user to introduce in the system a protein in a string format through a graphical user interface (see Fig. 6.2). This input is captured by an agent associated with the interface, which interacts with other agents to obtain a membership prediction for that specific query, using the multi-class support vector machine based algorithm previously presented. The constructed version uses the sequence similarity algorithm PSI-BLAST to get the profiles, discriminative models learned with a SVM algorithm and a neural network. Based on the prediction, it retrieves information concerning the protein superfamily and the respective class and fold from a local database.

The agent system follows the schema in Fig. 6.3, and its structure is composed by 3 varieties of agents:

1 AgentGUI - This agent is mainly responsible for connecting the user with the agent system. It has associated a user interface developed in the java language that allows submitting new jobs and visualizing the results. The agent starts by verifying the validity of the input, which must be a string composed only by characters belonging to the 20 possibilities from the standard amino acid alphabet. After a positive validation of the sequence format, it is transmitted to agent PredictionManager an order to start the prediction process for the introduced protein. In the final stage, AgentGUI receives a message from PredictionManager containing the prediction result. AgentGUI then gets the information associated with the predicted membership from a database and finally prints it in the user interface.

2 PredictionManager - As the name denotes this agent manages the prediction process, but it is also responsible for the execution of some calculations needed. When a new request arrives from AgentGUI, the PredictionManager starts by computing the Position Specific Scoring Matrix for the query protein, using for that the PSI-BLAST algorithm, performing after that the feature space projection. The dot product computation that follows, is then distributed among the Predictor agents available and registered with the JADE yellow pages service. After all agents Predictor finishing their tasks the PredictionManager joins all the subsolutions of the partitioned problem and continues with the next computational step. After receiving the decision values from each of the SVM models it finally gets the membership prediction from the neural network. The classification result is then sent to AgentGUI and it enters in a standby state waiting for a new job. The PredictionManager agent has a mechanism that allows submitting several jobs at the same time based on an id that is created using the date of the job submission. The id is used by all agents to keep track of the prediction process and to avoid erroneous combinations of results.

3 Predictor - This agent is a worker that executes calculations needed to get the decision values from the SVM discriminative models, namely the dot products needed to build the SVM kernels for the protein introduced by the user. The prediction system must have several agents of this kind according to the total number of processing units available.

4 Experiments

In this section are explained the most relevant experiments carried out to evaluate the efficiency of the classification models and the speedup gains introduced by the parallelization. The respective results are also presented and analyzed.

4.1 Learning efficiency

In order to choose the most suitable SVM algorithm, comparative measurements between the batch formulation LIBSVM [Chang and Lin 2004] and the incremental LASVM [Bordes et al. 2005], regarding the learning eficiency and complexity were made. This last algorithm was adapted in order to become able to deal with pre-computed kernels, since this was not an original trait.

A benchmark data set from SCOP previously used on remote homology detection [Weston et al. 2003] was reutilized to evaluate the classifiers. The data set has 7329 domains and was divided according to 54 families.

Remote, homology detection was simulated by considering all domains for each family as positive test examples and sequences outside the family but belonging to the same superfamily as positive train examples. Negative examples are from outside the positive sequences fold, and were randomly divided into train and test sets in the same ratio as the positive examples.

The profiles for the kernel were obtained with PSI-BLAST using 2 search rounds. Data was then projected to the feature space considering a subsequence of 5 amino acids and a threshold σ=7.5 as in [Kuang et al. 2005].

Receiver operating characteristic (ROC) was used to evaluate the quality of the SVM classifiers. A ROC curve consists in the plot of the true positives rate as a function of true negatives rate at varying decision thresholds, and expresses the ability of a model to correctly rank examples and separate distinct classes. A score can be obtained by calculating the area under the ROC curve (AUC) through the trapezoidal integration:

$$AUC = \sum \left\{ \left(1 - \varphi_i . \Delta\phi\right) + \frac{1}{2}\left[\Delta\left(1 - \varphi_i\right)\Delta\phi\right] \right\} \qquad (11)$$

, where sensitivity=1-φ and specificity=1- ϕ.

A good model has AUC=1, a random classifier is expressed by AUC~0,5 and the worst case comes when AUC=0.

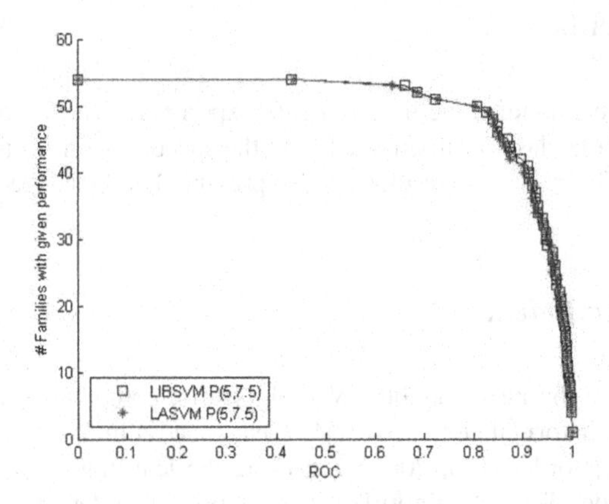

Fig. 6.4 Performance for LASVM and LIBSVM with the kernel profile (5, 7.5), where the plots express the cumulative number of families with at least a given ROC score.

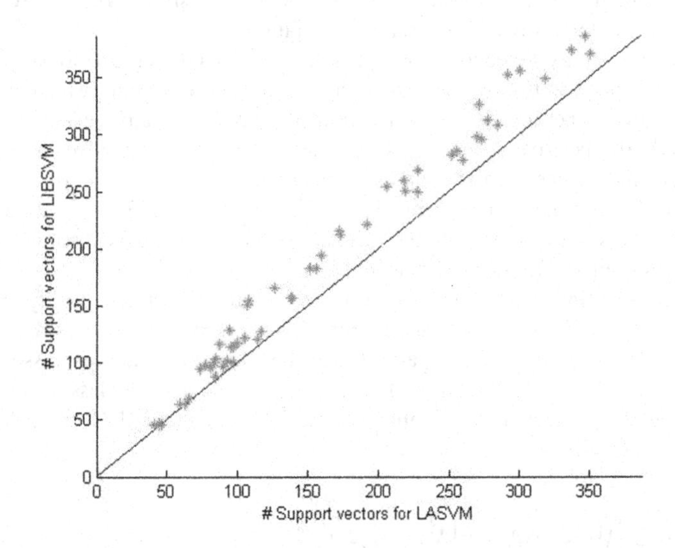

Fig. 6.5 Number of support vectors retained by the individual models trained using profile (5, 7.5). Each point is given by the values from LASVM plotted against the ones from LIBSVM, for each individual family.

LASVM and LIBSVM showed similar ability to learn (results can be checked in Fig. 6.4) and comparable processing time. Despite both algorithms achieved a mean ROC score of 0.92 for the 54 families, LASVM was able to build discriminative models with a reduced number of support vectors decreasing the learning complexity (as seen in Fig. 6.5), thus becoming the chosen SVM algorithm.

Train and test of the top neural network was accomplished according to a 10-fold cross-validation methodology, using the decision values from the SVMs. This last multiclass classifier achieved a global accuracy of 90% for the superfamily level.

4.2 Processing speed evaluation

The influence of the number of agents Predictor in the total processing time was also measured. This was done in a local machine with 4 cpus. The application was executed with 1, 2, 3, 4, 6 and 8 worker agents of type Predictor. Measurements were repeated 5 times for the same protein sequence and the average values used to calculate the speedup S_p for each case, given by:

$$S_p = \frac{T_1}{T_p} \tag{12}$$

, where T_1 is the execution time of the sequential algorithm and T_p is the execution time of the parallel algorithm with p worker agents of type Predictor.

Analyzing the speed improvements shown in Fig. 6.6, it is possible to see the benefits of concurrency, being achieved a maximum speedup of 1,7 when using 3 worker agents. It is important to keep in mind that the parallelization is implemented only in a portion of the prediction algorithm. As an example, the PSI-BLAST algorithm, that is responsible for a meaningful slice of the total processing time, runs sequentially so, the speedup values achieved by the whole system shall and do reflect that. After an almost linear speedup phase, the deterioration of the processing speed for more than 3 workers reveals from this point ahead that agents loose the capacity to further take advantage from the already saturated processing units introducing some extra overhead.

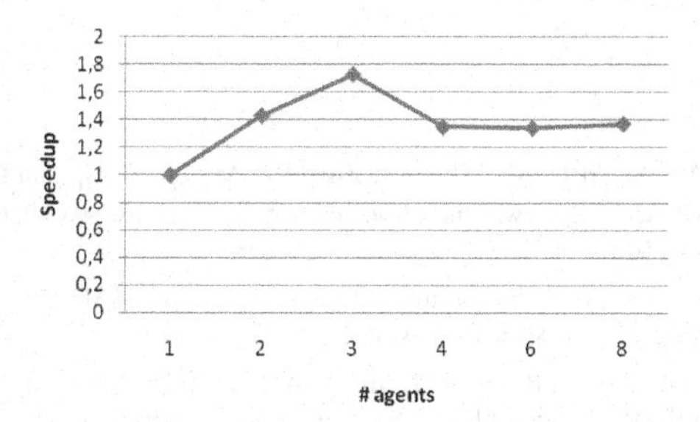

Fig. 6.6 Processing speed gain attained through the presented parallelization strategy. The plot represents the speedup as a function of the number of agents Predictor used in the platform.

5 Conclusions and future work

In this contribution we described a SVM based algorithm for the multi-class prediction of protein membership using a multi-agent system to parallelize the computations needed to get a prediction. A performance study conducted revealed the ability of the multi-class classifier scheme to accurately identify the superfamily of a given protein from its sequence alone and the considerable processing time reduction brought by the parallelization strategy implemented using the multi-agent system, even at a local level in a multi-core machine for a relatively small data set.

The parallelization follows a coarse grain strategy and was implemented considering the number of discriminative models involved in this specific problem. A finer tuning can further emphasize the benefits of parallelization in real situations where thousands of membership groups can be found.

The interest in creating tools that can take advantage of vulgar computational resources accessible virtually everywhere has taken us to extend the framework. Work is being done on two new application modules: one for parallelization of the train phase and another for automatic update of the discriminative models using the previous component. With the first new option it is intended to reduce the computation spent with jobs that commonly take several days to more reasonable amounts of time. On the other hand, because proteomics is a fast evolving field, where data available is constantly changing, the automatic actualization module is intended to keep up-to-date solutions with reduced human intervention and augmented system independence.

The base concepts presented here have also the potential to be used in data fusion solutions where extra information available in online databases can make the difference. This is another open door for future research.

Acknowledgments This work was supported by FCT – Fundação para a Ciência e a Tecnologia, under Project BIOINK – PTDC/EIA/71770/2006

References

Altschul SF, Madden TL, Schaffer AA, Zhang J, Zhang Z, Miller W, Lipman DJ (1997) Gapped BLAST and PSI-BLAST: A new generation of protein database search programs. Nucleic Acids Res 25: 3389-3402

Bellifemine F, Caire G, Trucco T, Rimassa G (2004) TILab SpA. JADE programmer's guide. http://jade.tilab.com/doc. Accessed 10 November 2008

Bellifemine F, Caire G, Trucco T, Rimassa G, Mungenast R (2005) TILab SpA. JADE Administrator's guide. http://jade.tilab.com/doc. Accessed 10 November 2008

Bordes A, Ertekin S, Weston J, Bottou L (2005) Fast Kernel Classifiers with Online and Active Learning. J Mach Learn Res 6: 1579-1619

Bradley AP (1997) The use of the area under the ROC curve in the evaluation of machine learning algorithms. Pattern Recognit 30(7): 1145-1159. doi: 10.1016/S0031-3203(96)00142-2

Caire G (2003) JADE Programming for beginners. TILab SpA. http://jade.tilab.com/doc. Accessed 10 November 2008

Chang C, Lin C (2004) LIBSVM: a Library for Support Vector Machines

Crammer K, Singer Y (2002) On the Learnability and Design of Output Codes for Multiclass Problems. J Mach Learn 47:201-233. doi: 10.1023/A:1013637720281

Crammer K, Singer.Y (2002) On the Algorithmic Implementation of Multi-class SVMs. J Mach Learn Res 2: 265-292

Dietterich TG, Bakiri G (1995) Solving Multiclass Learning Problems via Error-Correcting Output Codes. In AI Access Foundation. J Artif Int Res 2:263-286

Fawcett T (2006) An introduction to ROC analysis. Pattern Recognit Lett 27: 861-874

Hsu CW, Lin CJ (2002) A comparison of methods for multi-class support vector machines. IEEE Trans Neural Netw. 13 415-425. doi: 10.1109/72.991427

Jaakkola T, Diekhans M, Haussler D (1999) Using the Fisher Kernel Method to Detect Remote Protein Homologies. Proc Int Conf Intell Syst Mol Biol

Joachims T, Finley T, Yu C (2009) Cutting-plane training of structural SVMs. J Mach Learn 77. doi: 10.1007/s10994-009-5108-8

Krogh A, Brown M, Mian I, Sjolander K, Haussler D (1994) Hidden markov models in computational biology: Applications to protein modeling. J Mol Biol 235: 1501-1531. doi: 10.1006/jmbi.1994.1104

Kuang R, Ie E, Wang K, Wang K, Siddiqi M, Freund Y, Leslie C (2005) Profile-based string kernels for remote homology detection and motif extraction. J Bioinform Comput Biol 3:527-550. doi: 10.1142/S021972000500120X

Leslie C, Eskin E, Cohen A, Weston J, Noble, W (2004) Mismatch string kernels for discriminative protein classification. Bioinformatics 20:467-476. doi: 10.1093/bioinformatics/btg431

Leslie C, Eskin E, Noble W (2002) The spectrum kernel: a string kernel for SVM protein classification. In: Proc Pac Symp Biocomput 7, 564-575

Melvin I, Ie E, Kuang R, Weston J, Noble W, Leslie C (2007) Svm-fold: a tool for discriminative multi-class protein fold and superfamily recognition. BMC Bioinforma 8(4). doi: 10.1186/1471-2105-8-S4-S2

Murzin AG, Brenner SE, Hubbard T, Chothia C (1995) SCOP: A structural classification of proteins database for the investigation of sequences and structure. J. Mol. Biol. 247: 536-540. doi: 10.1006/jmbi.1995.0159

Rifkin R, Mukherjee S, Tamayo P, Ramaswamy S, Yeang C, Angelo M, Reich M, Poggio T, Lander E, Golub T, Mesirov, J (2003) An analytical method for multi-class molecular cancer classification. Society for Industry and Applied Mathematics Reviews 45: 706-723

Vapnik V N (1998) Statistical learning theory. Adaptive and learning systems for signal processing, communications, and control. Wiley, New York

Weston J, Leslie C, Zhou D, Elisseeff A, Noble W (2003) Semi-Supervised Protein Classification using Cluster Kernels.In: Adv Neural Inf Process Syst 17. doi: 10.1093/bioinformatics/bti497

Modeling and Control of a Dragonfly-Like Robot

Micael S. Couceiro[1], N. M. Fonseca Ferreira[1] and J.A. Tenreiro Machado[2]

[1]Institute of Engineering of Coimbra

Rua Pedro Nunes - Quinta da Nora, 3030-199 Coimbra, Portugal

(email: micaelcouceiro@gmail.com; nunomig@isec.pt)

[2]Institute of Engineering of Porto

Rua Dr. António Bernardino de Almeida, 4200-072 Porto, Portugal

(email: jtm@isep.ipp.pt)

Abstract Dragonflies demonstrate unique and superior flight performances than most of the other insect species and birds. They are equipped with two pairs of independently controlled wings granting an unmatchable flying performance and robustness.
In this paper it is studied the dynamics of a dragonfly-inspired robot. The system performance is analyzed in terms of time response and robustness. The development of computational simulation based on the dynamics of the robotic dragonfly allows the test of different control algorithms. We study different movement, the dynamics and the level of dexterity in wing motion of the dragonfly.
The results are positive for the construction of flying platforms that effectively mimic the kinematics and dynamics of dragonflies and potentially exhibit superior flight performance than existing flying platforms.

1 Introduction

The study of dynamic models based on insects is becoming popular and shows results that may be considered very close to reality [Schenato et al. 2001] and [Wang 2005]. One of the models under study is based on the dragonfly [Tamai et al. 2007] because it is considered a major challenge in terms of dynamics. Recent studies show that the aerodynamics of dragonflies is unstable because they use a flying method radically different from steady or quasi-steady flight that occurs in aircrafts and flapping or gliding birds [Kesel 2000]. This unsteady aerodynamic has not received proper attention due to the inherent level of complexity.

Recently, technological advances allow the construction of robotic systems that are able to perform tasks of some complexity. In the past, there were significant advances in robotics, artificial intelligence and other areas, allowing the implementation of biologically inspired robots [Cohen and Breazeal 2003]. Therefore,

A. Madureira et al. (eds.), *Computational Intelligence for Engineering Systems: Emergent Applications*, Intelligent Systems, Control and Automation: Science and Engineering 46,
DOI 10.1007/978-94-007-0093-2_7,

researchers are investing in reverse engineering based on the characteristics of animals. The progress of technology resulted in machines that can recognize facial expressions, understand speech and perform movements very similar to living beings.

Some interesting examples are spiders [Vallidis 2008], snakes [Spranklin 2006], insects [Lauder 2001] and birds [Ellison 2006, Couceiro et al. 2009a]. They all require an extensive study of both the physical and the behavioral aspect of real animals.

The paper is organized as follow. Section two presents the state of the art in the area. Sections three and four provide an overview of the physical structure and the kinematics of the dragonfly, respectively. Sections five and six describe the dragonfly dynamics and the flight process implemented through the proposed model. Section seven and eight develop the dynamical analysis and the control algorithms, respectively. Finally, section nine outlines the main conclusions.

2 State of the Art

Inspired by the unique characteristics of animals, researchers have placed a great emphasis on the development of biological robots. This chapter addresses the studies and previous work done in this area focusing on the development of robots inspired in flying animals.

The flight of insects has been an interesting subject of, at least, half a century, but serious attempts to recreate it are much more recent [Zbikowski 2005]. Aircraft designers have been interested in increasing the morphic capabilities of wings and this area received a major boost in 1996, when the Defense Advanced Research Projects Agency of the U.S. (DARPA) launched a MAV of three years in order to create a flying platform with less than 15 centimeters long for surveillance and reconnaissance.

Some other biological inspired platforms have been developed such as the *Dragonfly* from *Wow Wee!*. The *Dragonfly* toy was developed in 2007 and it is controlled by a radio transmitter. It looks like a dragonfly with a wingspan of 40.6 centimeters, with a lightweight body and strong double wings. As the dragonfly beats the wings to fly it does not need a propeller to generate a thrust force. It only uses a propeller in the tail to move left or right.

In 2008 a robotic platform inspired by the flight of birds was developed at ISEC (Institute of Engineering of Coimbra). SIRB (*Simulation and Implementation of a Robotic Bird*) was built based on the results obtained using a simulator developed in *Matlab* [Couceiro et al. 2008].

While the developments of robotic platforms described above are a positive step in the production of new biologically inspired flying robots, there is a subarea that does not have the proper attention of researchers: the control and autonomous navigation of robots.

Some studies have been appearing on the area of autonomous navigation of flying robots, studying new techniques of odometry and vision [Iida and Lambrinos 2001]. Fumiya Iida developed control algorithms with the Reichard model conducting experiments in an autonomous flying airship robot in an unstructured environment [Iida 2001].

The control of flying robots, even if not inspired in flying animals, represents a high level of complexity. Sukon Puntunan and Manukid Parnichkun [Pununan and Parnichkun 2006] compared the classical *PID* with a self-tuning *PID* algorithm for the control a small helicopter. The results obtained with the self-tuning *PID* proved that this type of control offers a better performance than the classical *PID*. However, it was possible to observe some relatively high overshoots in the system response.

In this paper we address another control and optimization methods comparing the results obtained in order to make the system steadier and, thereby, obtaining a better performance.

3 The Kinematics of the Dragonfly

The dragonfly model is being studied by some researchers due to the unique juggling maneuvers of this creature. Jane Wang [Wang 2005] developed a set of equations based on a real model of a dragonfly by watching its flight in laboratory.

The objective in defining the geometry is to develop a physical model that can be mathematically described as being comparable to the actual real dragonfly. Based on some works already developed in this area, and performing a geometric analysis of the dragonfly, it was possible to reach a relatively simple model with a high-quality response when comparing to what it is observed in nature.

As we can see, the major difference between the geometry of two-winged animals (e.g., birds) and the geometry of the dragonfly are reflected in two pairs of wings.

Similarly to birds, the dragonfly also has several movements and flying styles. The flight capabilities of dragonflies are prodigious. In addition to the individual states of take-off, gliding and flapping, this last one is divided into four different styles due to the two pairs of wings: counter-stroking (where the front and rear wings beat with a delay of 180 degrees), phased-stroking (in which the wings beat with a difference of 90 degrees), synchronized-stroking (in which the four wings are synchronized as a single pair of wings), and gliding such as occurs in large birds. We will give special attention to the most common style in which the two pairs of wings of the dragonfly beat with a delay of 180 degrees (counter-stroking) that will be explained in the sequel.

Based on the geometry, and following an analysis of the multi-link model, we estimated the location of every joint in the robot and obtained the kinematic model represented in Fig. 7.1.

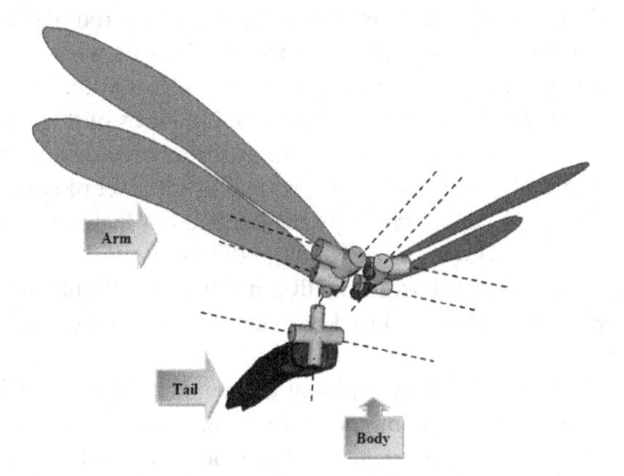

Fig. 7.1 Kinematic structure of the system.

The tail and each pair of wings have the same degrees of freedom (rotational) found in other flying models such as birds. The wings will be treated as a flexible link, similarly to what is seen in the nature, for minimizing the area of the wing when on the downward movement. This structure will provide a good mobility, making it a total of ten controllable links.

4 The Dynamics of the Dragonfly

The dragonfly dynamics is somehow similar to other flying creatures such as birds [Couceiro et al. 2009b] and, consequently, the same equations may be considered. Nevertheless, when it comes to the flapping flight, the dragonfly takes a great advantage over birds and other two-winged creatures (Fig. 7.2).

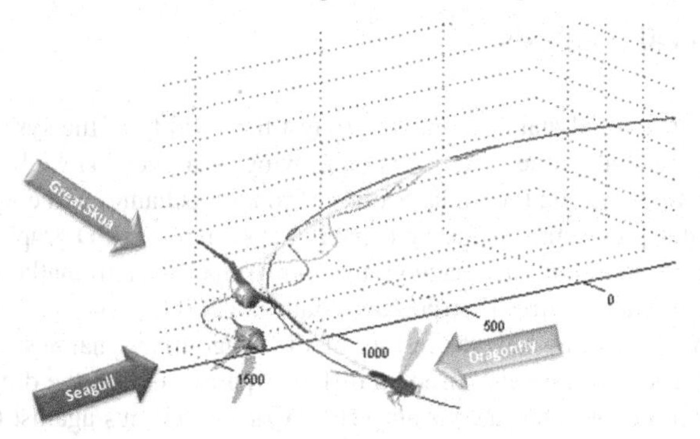

Fig. 7.2 Chart obtained through the developed simulator that shows the difference between the trajectory accomplished by a great skua (very large bird), a seagull (large bird) and a dragonfly. The stability of this last one when compared to the others is undeniable.

Recent studies reveal that dragonflies use a complex aerodynamics to fly, differently from aircrafts and large birds. A dragonfly flaps its wings to create a whirlwind of air that is controlled and used to provide lift. On the other hand, planes depend on good air flow over the top and bottom surfaces of their wings. For these machines the turbulence can be fatal. There are other creatures with a mechanism similar to the flight of the dragonfly, but with a higher level of complexity, such as the hummingbird, that can surprisingly manipulate the feathers of the wings during the rapid flapping. However, the study of dragonfly flight shows that it can be as efficient as the hummingbird but with a much easier flight system. More than 200 million years of evolution provide evidences of a successful and infallible aerodynamics.

The two pairs of wings allow different independent flight techniques (as mentioned above) and the most common style is the counter-stroking. This type of flight allows that, when a pair of wings beats down creating a vortex of air, the other pair, which is still down, captures the energy of that vortex. Therefore, the air flow over the surface of the wings of the dragonfly has a much higher rate along the bottom of the wing creating more lift. In other words, the different states of flight, downstroke and upstroke, are indistinguishable creating an almost steady force positive to the movement and contrary to the weight. Nevertheless, applying this principle to the development of flying platforms is complex because the effect has to be simple and predictable. Less than ten years ago, people saw the flows generated by the insects as something uncontrollable. The turbulence was, and still is, often seen as something undesirable, causing failures in the turbines of the aircrafts and reducing their effectiveness. In the case of the rotor of helicopters, the blades sometimes fail because each blade is continuously affected by the turbulence generated by the preceding blade, causing vibrations that may weaken the metal. However, for the dragonfly, this type of flight is something natural and extremely efficient as we shall see in the next section.

5 Dynamical Analysis

We have undertaken a dynamical analysis to test the validity of the system model. In order to easily change the parameters (*e.g.*, wing area, weight) we built a computer program highlighting the fundamentals of robot mechanics and control.

The computer programs emphasize capabilities such as the 3D graphical simulation and the programming language giving some importance to mathematical aspects of modeling and control [Ferreira and Machado 2000].

We start by presenting several results of the dragonfly dynamics around the gliding flight. These results are based on different parameters of the dragonfly. In each simulation the wind has a constant velocity of $v = 5.0$ m/s against the movement of the dragonfly that has an initial velocity of $v_0 = 3.0$ m/s. We change the weight and the area of the wing parameters in order to analyze the dragonfly dynamics. The initial parameters are a total weight of $m = 10^{-3}$ kg and the wing an area of $S = 10^{-4}$ m^2.

For the dragonfly to fly in a straight line, without flapping its wings, it is needed a continuously changing of the angle of attack (alpha) to keep a vertical resulting force equal to zero. The angle of attack will then increase the lift and the drag forces. A higher drag force results in the reduction of the velocity. This process stops when the velocity reaches zero since we do not want the dragonfly to be dragged by the wind.

In the following experiments that can be seen in Fig. 7.3-7.4 we will change the mass and the wing area in increments of 25% and 10% of the initial parameters, respectively.

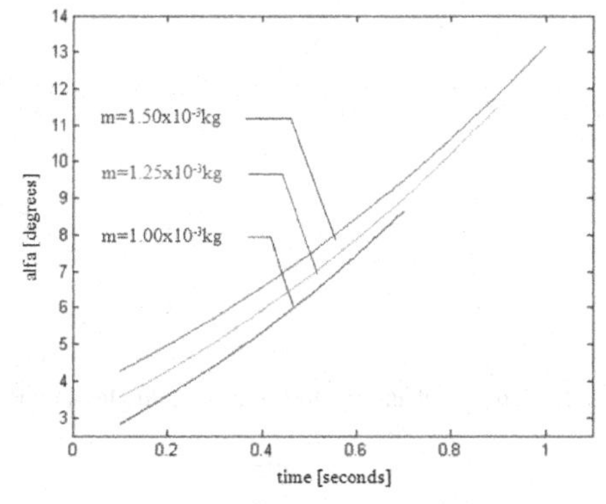

Fig. 7.3 Dragonfly gliding straight - changing the weight. Angle of attack versus time.

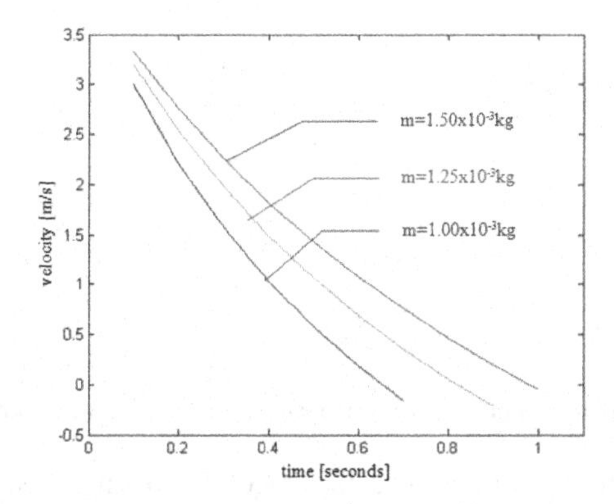

Fig. 7.4 Dragonfly gliding straight - changing the weight. Velocity versus time.

As we can see, increasing the weight requires a higher angle of attack in order to fly. The dragonfly keeps gliding for a short amount of time when compared to large birds. Despite the weight that is also well below the weight of the large flying creatures, like soaring birds, the area of the wings does not allow gliding for a long time. Obviously, the dragonfly, like all insects or small birds, does not have the same ability to glide as a large bird.

An interesting aspect is the fact that by increasing the weight of the dragonfly it can glide longer. This can easily be explained: if you throw a feather against the wind it will not go as far as if you throw a stone. As we increase the weight of the dragonfly we are giving it the chance to fight against the wind more easily; however, we are also ensuring that it needs a higher angle of attack of the wings which, on the other hand, will eventually reduce the speed anyway.

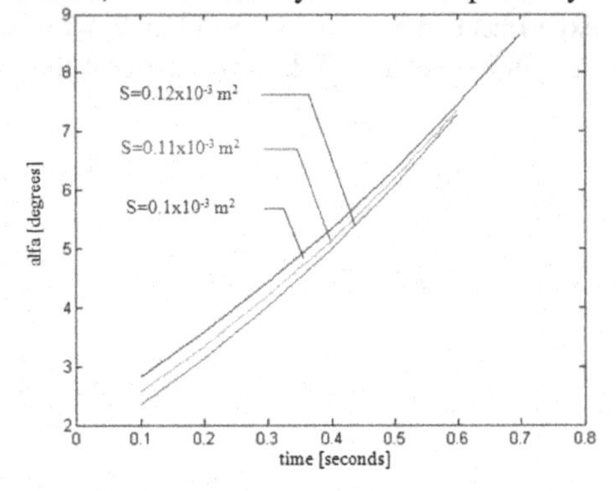

Fig. 7.5 Dragonfly gliding straight - changing the wing area. Angle of attack versus time.

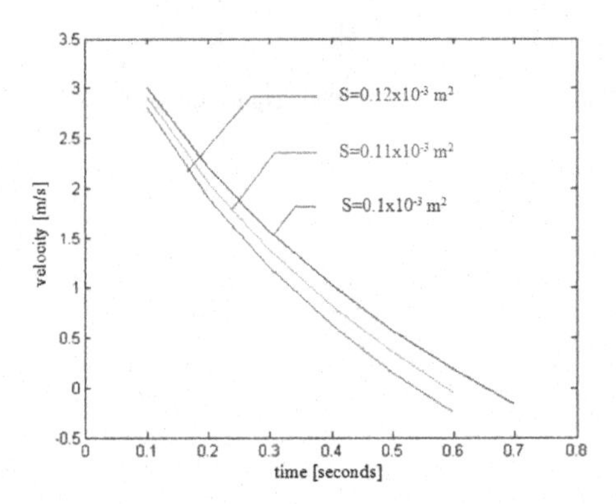

Fig. 7.6 Dragonfly gliding straight - changing the wing area. Velocity versus time.

By increasing the area of the wings the dragonfly does not need to significantly increase the angle of attack because it can keep gliding more easily (Fig. 7.5 and 7.6).

Birds, particularly large ones, adopt this technique much more frequently than insects do. Nevertheless, insects also use it, although not with the purpose of saving energy, since the difference is not relevant, but to accomplish some specific maneuvers.

The second experiment (Fig. 7.7-7.10) shows the horizontal (v_x) and vertical (v_z) velocities of the bird as well as the vertical distance obtained when the bird is gliding down a vertical distance of 5.0 meters, when considering a fixed angle of attack in both wings.

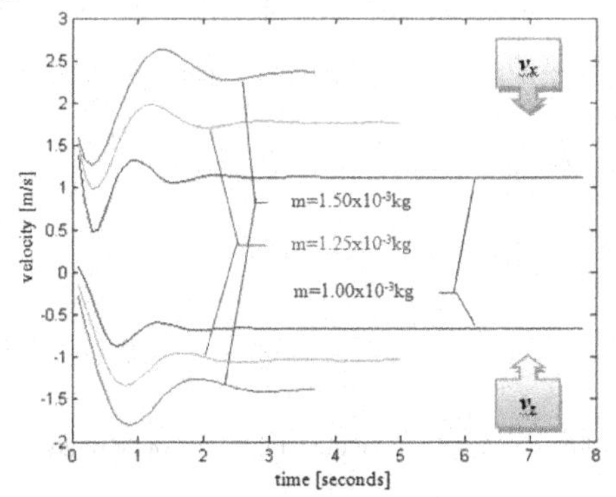

Fig. 7.7 Dragonfly gliding down - changing the weight. Angle of attack versus time.

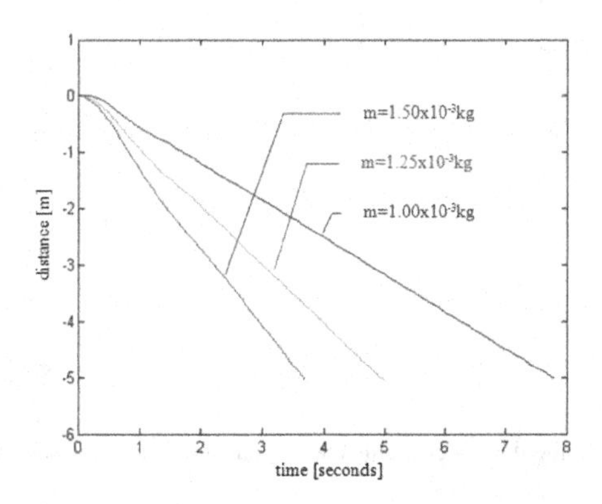

Fig. 7.8 Dragonfly gliding down - changing the weight. Distance versus time.

It is obvious that, when we increase the weight of the dragonfly, it reaches the desired vertical distance faster. However, there is a ubiquitous aspect that must be emphasized: the movement is much more linear than the movement of larger creatures such as birds. The reason is the relation between the area of the wings and the weight.

Let us compare the flight with the one of a large bird: while the wings of the dragonfly are, let us suppose, 100 times smaller than the wings of the bird, the weight of the dragonfly is about 400 times smaller. By doing this imbalance in the weight/area of the wings we assure that the flying movement is more linear. Based on what we just said and taking into account the large difference between the weight/area of the wings of the dragonfly, if we increase the area of the wing even more then the movement will be even more linear. We can confirm the idea in Fig. 7.9 and 7.10.

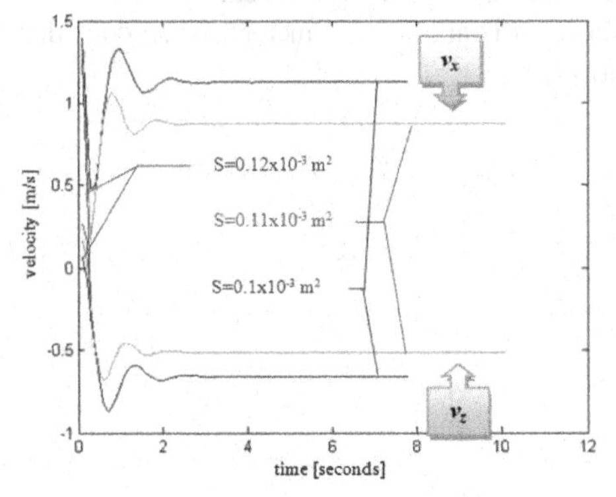

Fig. 7.9 Dragonfly gliding down - changing the wing area. Angle of attack versus time.

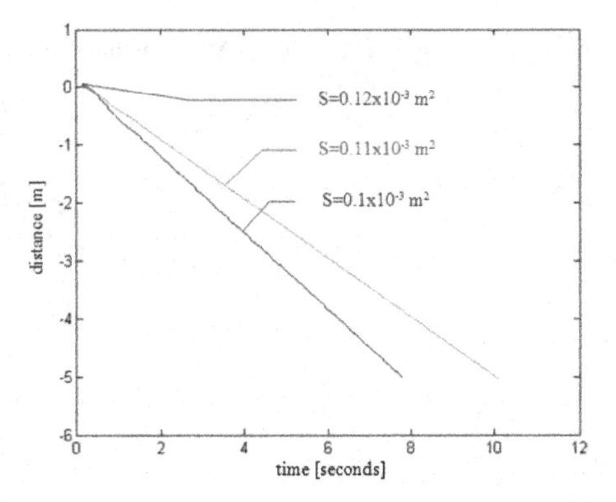

Fig. 7.10 Dragonfly gliding down - changing the wing area. Distance versus time.

Nevertheless, this relationship is not as straight as it seems in the previous charts. It is true that increasing the area of the wings by 10% the movement becomes more linear and it can eventually perform the desired trajectory smoothly and with a lower speed. However, increasing the area over 10% the dragonfly cannot achieve the desired position. This is due to the fact that the size of the wings are so large, when compared to the weight, that the drag caused by the wings is too high so that the resultant force in x-axis reaches zero.

This shows that the relationship weight/area of the wings of the dragonfly is ideal, and that manipulating this relationship can eventually have unexpected results and may compromise the good efficiency of the dragonfly flight.

We will now analyze the flapping flight of the dragonfly to understand how it works in order to implement a control algorithm. The analysis of the flapping flight is not as simple as for the case of the gliding flight. In the next experiment, we must note that our first priority is to fly in a straight line.

Following a similar line of thought of the gliding flight we change the weight and wing area. Fig. 7.11 and 7.12 show how the velocities and vertical distance react while changing the bird weight.

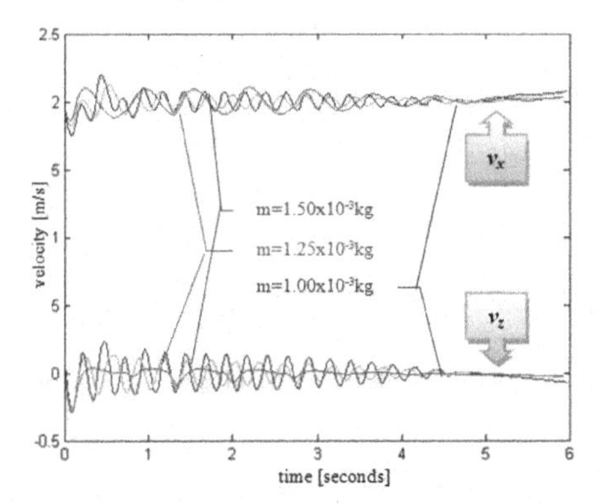

Fig. 7.11 Dragonfly flapping straight - changing the weight. Angle of attack versus time.

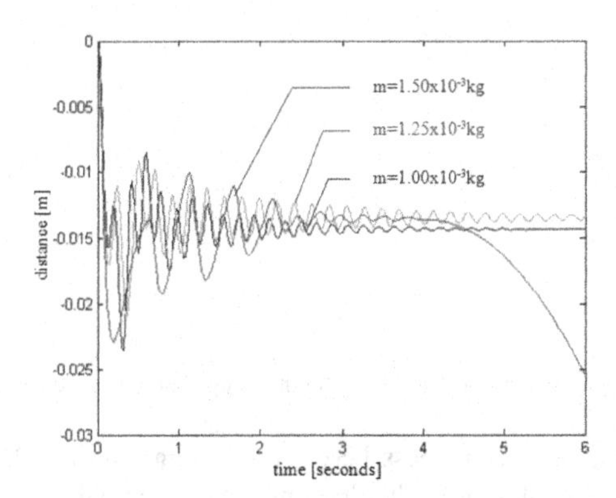

Fig. 7.12 Dragonfly flapping straight - changing the weight. Distance versus time.

The previous figures shows that the dragonfly can maintain a very straight trajectory except for a weight 50% higher, because as it begins to slightly lose some altitude. However, the flight starts with an initial velocity v_0 = 2 m/s and remains near this value even with the significant increase in the weight.

It is easy to understand that, if we increase the area of the wings of the dragonfly (Fig. 7.12 and 7.13) then the flapping wings response will be enhanced. This effect is opposed to the previous experiment, where the significant increase of the area of the wing brought some inconvenient in the gliding flight, because of the lack of thrust force. A larger area of the wings means a smaller settling time of the dragonfly velocity as can be easily seen in Fig. 7.13.

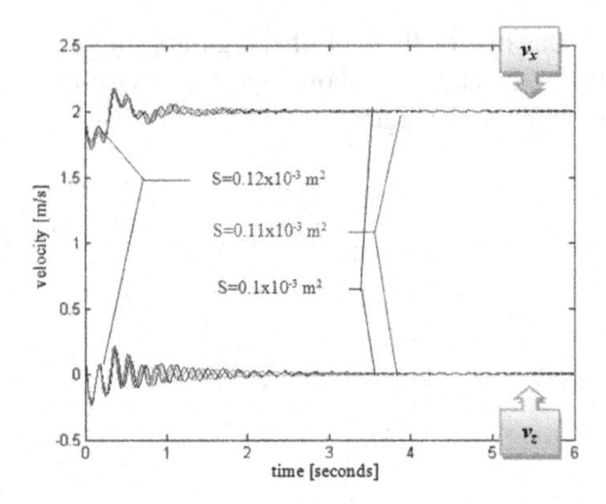

Fig. 7.12 Dragonfly flapping straight - changing the wing area. Angle of attack versus time.

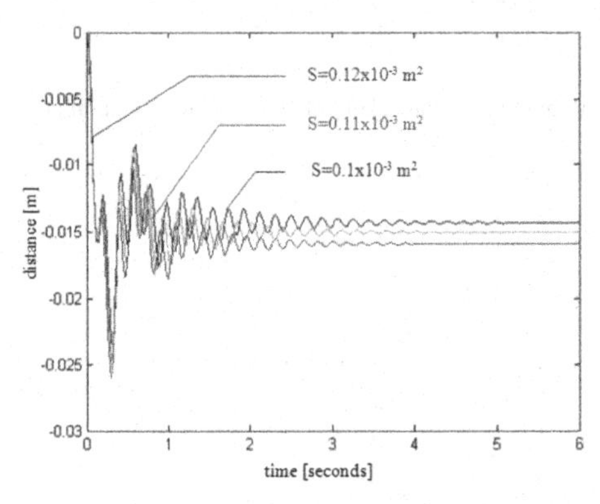

Fig. 7.13 Dragonfly flapping straight - changing the wing area. Distance versus time.

The difference of effectiveness between the dragonfly and large birds mainly focuses on the flight stability. The dragonfly can eventually overcome variations in the parameters (e.g., weight, area of the wings) more easily than birds and other two winged creatures. The dragonfly maintains a regular wing-beat of 3.0 to 5.0 flaps/s (depending on the weight and wing area) not making use of the gliding flight such as large birds do. The experiments in the next section with the optimized controllers will give us a better understanding about the real stability and performance of the dragonfly flight.

6 Controller Performances

In this section we develop several experiments for comparing the performances of the *FO* (Fractional Order) *PID* algorithms [Ferreira et al. 2002, Couceiro et al. 2009c].

FO controllers are algorithms whose dynamic behavior is described thorough differential equations of non integer order. Contrary to the classical *PID*, where we have three gains to adjust, the *FO PID*, also known as $PI^{\lambda}D^{\mu}$ ($0 < \lambda$, $\mu \leq 1$), has five tuning parameters, including the derivative and the integral orders to improve de design flexibility.

The mathematical definition of a derivative of fractional order α has been the subject of several different approaches such as the Laplace: The Grünwald-Letnikov definition is perhaps the best suited for designing directly discrete time algorithms.

$$D^{\alpha}(x(t)) = \lim_{k \to 0}\left[\frac{1}{h^{\alpha}}\sum_{k=1}^{\infty}\binom{\alpha}{k}x(t-kh)\right] \tag{1}$$

$$\binom{\alpha}{k} = \frac{(-1)^k\Gamma(\alpha+1)}{\Gamma(k+1)\Gamma(\alpha-k+1)} \tag{2}$$

where Γ is the gamma function and h is the time increment.

For the implementation of the $PI^{\lambda}D^{\mu}$ given by:

$$G_c(s) = K\left(1 + \frac{1}{T_i s^{\lambda}} + T_d s^{\mu}\right) \tag{3}$$

we adopt a 4th-order discrete-time Pade approximation.

To tune the controllers' parameters we used a medium scale *Gradient Descent* method with 200 maximum iterations. To find a local minimum of a function of the position error using gradient descent, one takes steps proportional to the negative of the gradient (or the approximate gradient) of the function at the current point.

The first attempt to control our system will be changing the wing speed velocity, angle of attack and tail rotations accordingly with the cartesian position error.

In order to study the system response to perturbations, during the contact we apply, separately, rectangular pulses, at the references. Therefore, the trajectory used to optimize the controllers consists in a straight line flight with a velocity of $v_x = 1$ m/s during the first 20 seconds. The dragonfly will then need to instantaneously achieve a velocity of $v_x = 3$ m/s. Finally, 20 seconds later, the system will instantaneously reduce is velocity to $v_x = 1$ m/s again.

In this optimization it is unnecessary the use of a controller in the *y*-axis since there will be no movement in this axis; therefore, we will ignore it for now.

Let us then compare the *PID* and $PI^{\lambda}D^{\mu}$ controllers. Under the last conditions we obtained the *PID* and $PI^{\lambda}D^{\mu}$ controller parameters depicted in Table 7.1.

Table 7.1 PID and PIλDμ controller parameters.

-	K_{pX}	K_{iX}	K_{dX}	μ_X	λ_X	K_{pZ}	K_{iZ}	K_{dZ}	μ_Z	λ_Z
PID	60	0	13	-	-	125	65	25	-	-
$PI^{\lambda}D^{\mu}$	36	0	5	0.85	0.9	106	70	25	0.8	0.6

To analyze more clearly the dynamical response to the step perturbation we subtract the dynamic response without perturbation. Fig. 7.14 depict the system response under the action of the *PID* and $PI^{\lambda}D^{\mu}$ algorithms.

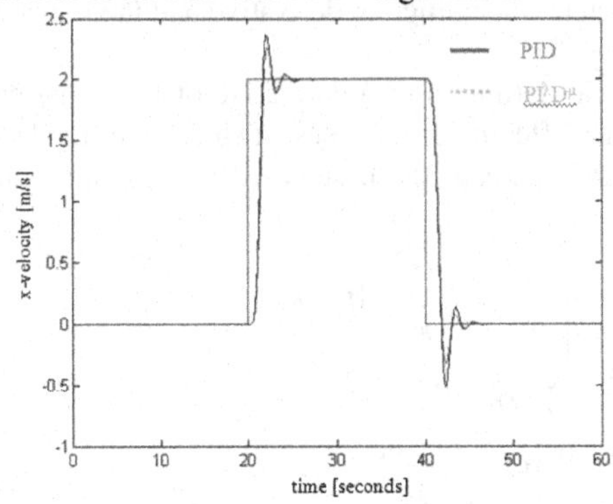

Fig. 7.14 Time response of the system under the action of the PID and PIλDμ controllers.

Table 7.2 compares the time response characteristics of the integer and the fractional *PID* controllers, namely the percent overshoot *PO*, the rise time t_r, the peak time t_p and the settling time t_s.

Table 7.2 Time response parameters of the system under the action of the PID and $PI^{\lambda}D^{\mu}$ controllers

	PO(%)	t_r	t_p	t_s
PID	18.25	20.74	21.16	25.52
$PI^{\lambda}D^{\mu}$	13.16	20.86	21.26	25.58

We can see that the *FO* algorithm takes advantage in the time response analysis, with a significant reduction of the overshoot.

7 Conclusion

The functionalities presented in this work are implemented in a simulation platform. We obtain satisfactory results proving that the development of the kinematical and dynamic model can lead to the implementation of an artificial machine with a behavior close to the dragonfly.

The design methodology and implementation can be deemed successful in this project. By obtaining a balance between physical modeling and the objective of animation, a strong advance in the system design has been achieved. Despite all simplifications, our model is still incomplete, and further research needs to be conducted to explore additional abstractions.

References

Bar-Cohen Y, Breazeal C (2003) Biologically-Inspired Intelligente Robots, SPIE Press, Vol. PM122

Couceiro M, Ferreira N, Machado J (2009c) "Application of Fractional Algorithms in the Control of a Robotic Bird". Journal of Comunications in Nonlinear Science and Numerical Simulation-Special Issue, Elsevier

Couceiro M, Figueiredo C, Ferreira N, Machado J (2008) Simulation of a robotic bird, Fractional Differentiation and its Applications. Ankara, Turkey

Couceiro M, Figueiredo C, Ferreira N, Machado J (2009a) Biological inspired flying robot, Proceedings of IDETC/CIE 2009 ASME 2009 International Design Engineering Technical Conferences & Computers and Information in Engineering Conference August 30 - September 2, San Diego

Couceiro M, Figueiredo C, Ferreira N, Machado J (2009b) The Dynamic Modeling of a Bird Robot, 9th Conference on Autonomous Robot Systems and Competitions, Robotica 2009, Castelo Branco, Portugal

Ellison A (2006) Cybird, Product Review in FlyingToys, p. 92

Ferreira N, Barbosa J, Machado J (2002) Fractional- Order Position/Force Control of Mechanical Manipulators. Proceedings of CIFA'02, Conférence Internationale Francophone d´Automatique 8-10 July, Nantes, France

Ferreira N, Machado J (2000) RobLib: An educational program for analysis of robots, Proceedings of Controlo 2000, 4th Portuguese Conference on Automatic Control, pg. 406-411 4-6 Oct, Guimaraes – Portugal

Iida F (2001). Goal-Directed Navigation of an Autonomous Flying Robot Using Biologically Inspired Cheap Vision, Proceedings of the 32nd International Symposium on Robotics (ISR), 1404-1409

Iida F, Lambrinos D (2001) Navigation in an Autonomous Flying Robot by Using a Biologically Inspired Visual Odometer, Proc. SPIE Vol. 4196, p. 86-97

Kesel A (2000) Aerodynamic Characteristics of Dragonfly Wing Sections Compared with Technical Aerofoils, Journal of Experimental Biology, Vol 203, Issue 20 3125-3135

Lauder G (2006) Flight of the Robofly, Nature, Macmillan Publishing Ltd., 0028-0836

Pununan S, Parnichkun M (2006) Online Self-Tuning Precompensation for a PID Heading Control of a Flying Robot, International Journal of Advanced Robotic Systems

Schenato L, Deng X, Wu W, Sastry S (2001) Virtual Insect Flight Simulator (VIFS): A Software Testbed for Insect Flight, IEEE Int. Conf. Robotics and Automation, Seoul, Korea

Spranklin B (2006) Design, Analysis, and Fabrication of a Snake-Inspired Robot with a Rectilinear Gait, Thesis at MS University of Maryland, College Park, 2006, 218 pages AAT 1436363

Tamai M, Wang Z, Rajagopalan G, Hu H (2007) Aerodynamic Performance of a Corrugated

Dragonfly Airfoil Compared with Smooth Airfoils at Low Reynolds Numbers, 45th AIAA Aerospace Sciences Meeting and Exhibit, Reno, Nevada

Vallidis N (2008) A Hexapod Robot and Novel Training Approach for Artificial Neural Networks, CiteSeer

Wang Z. (2005) Dissecting Insect Flight, Annu. Rev. Fluid Mech, 183-210

Zbikowski R (2005) Fly Like a Fly, IEEE Spectrum, 42(11), 46–51

Emotion Based Control of Reasoning and Decision Making

Luis Morgado[1] **and Graça Gaspar**[2]

[1]ISEL, Rua Conselheiro Emídio Navarro, 1949-014 Lisboa, Portugal, *(email: lm@deetc.isel.pt)*
[2]LabMAg - FCUL, Campo Grande, 1749-016 Lisboa, Portugal, *(email: gg@di.fc.ul.pt)*

Abstract In real-world domains, where uncertainty and dynamism are pervasive and time and resources are limited, reasoning and decision-making processes raise important problems related both to adaptive ability and to the computational complexity of the underlying cognitive mechanisms. In this context the integration between emotion based mechanisms and cognitive mechanisms can play a key role to support the development of intelligent agents able of effective behavior under real-time resource-bounded conditions. In this paper we address those issues under the framework of the agent flow model, an agent model where emotion and cognition are modeled as two integrated aspects of intelligent behavior and where affective-emotional mechanisms are used to support adaptability and to focus the reasoning and deliberation mechanisms to cope with their computational complexity.

1 Introduction

Relevant theoretical and experimental work (e.g. Damásio 2000, LeDoux 2000) has demonstrated the fundamental role that emotion plays in reasoning and decision-making. For example, experimental results reported by Damásio [Damásio 2000] indicate that a selective reduction of emotion is at least as prejudicial for rationality as excessive emotion, and Gray et al. [2002] reported neural evidence for a strong highly constrained form of emotion-cognition interaction, with loss of functional specialization, indicating that emotion and higher cognition can be truly integrated. On the other hand, the importance of emotional phenomena in learning and adaptive behavior is also well documented (e.g. LeDoux 2000). This evidence of an encompassing role of emotion in cognitive activity remains largely unexplored in cognitive models for intelligent agents. We explore this line of evidence by proposing that a symbiotic integration between emotion and cognition is a key aspect for the implementation of agents capable of intelligent behavior in complex, dynamic and uncertain conditions, typical of real environments.

However, two main problems are recognized underlying this emotion-based approach: (i) the tightly intertwined relation between emotion and cognition, which is hardly compatible with patching emotional phenomena as an addition to

A. Madureira et al. (eds.), *Computational Intelligence for Engineering Systems: Emergent Applications*, Intelligent Systems, Control and Automation: Science and Engineering 46,
DOI 10.1007/978-94-007-0093-2_8,

the cognitive mechanisms of an agent [Damásio 2000]; (ii) the dynamic and continuous nature of emotional phenomena, which is highly constrained by the classical notion of a discrete emotional state and its assessment via verbal labels [Scherer 2000]. In our view, to address these issues we must go beyond the classical separation between emotion and cognition and recognize their symbiotic relation. That is, emotion is a result of cognitive activity and cognitive activity is modulated by emotion, in a dynamic process that unfolds through time according to agent-environment interaction. We concretized this view by developing an agent model where emotion and cognition are modeled as two integrated aspects of intelligent behavior. Three main aspects are involved in this relation between emotion and cognition: (i) the regulation of cognitive activity due to the present achievement conditions; (ii) the modulation of the changes in the cognitive structure due to past experiences (i.e. the formation of emotional memories); (iii) the communicative role of emotional phenomena. In this paper we will focus on the first two aspects, that is, the control of high level cognitive processes, such as reasoning, to cope with their computational complexity and the control of memory mechanisms to enhance the adaptation and decision-making under time-limited conditions.

The paper is organized as follows: in sections 2 and 3 we present an overview of the emotion and agent models that support the proposed approach; in section 4 we further detail the agent model regarding adaptive reasoning; in section 5 we address emotional memory formation and activation and its use in decision-making; in section 6 we establish comparisons with related work and draw some conclusions.

2 Modeling Artificial Emotion

The study of emotional phenomena, particularly the development of emotion models for intelligent agent implementation has been mainly based on a perspective of emotion as a human phenomenon, possibly shared by some animals in the evolutionary continuum, according to the existence of specific brain structures. Cognitive appraisal models are a clear example of this approach, but even physiologically based models are inspired by human physiology and behavior. This anthropomorphic approach has an important drawback. Due to the tightly intertwined relation between emotion and cognition [Arzi-Gonczarowski 2002] emotional phenomena in this context are highly complex with multiple emotion blends. In fact phenomenological observations indicate that the complexity of emotional phenomena has its highest in humans [Hebb and Thompson 1979]. From this point of view, we can understand why multiple emotion models and theories coexist and their difficulty to characterize emotion in a concise way, especially in what relates to the processes underlying emotional phenomena.

Alternative views have been proposed by some authors [e.g. Scherer 2000, Carver and Scheier 2002] where these dynamic aspects are considered. However they maintain a typical commitment to the anthropomorphic view of emotion, which leads to complexity, brittleness and lack of flexibility in agent design and implementation, especially if we consider agents of different kinds and levels of complexity.

On the other hand, in the evolutionary continuum there is no evidence of a discontinuity in what regards the existence of emotional phenomena. On the contrary, it is well known that some simple organisms, even unicellular organisms, can present remarkable behaviors for organisms without nervous system, which from an observer point of view are easily classified as emotional [Staddon 2001].

Although almost unexplored, these observations give rise to the possibility that emotional phenomena are rooted not on cognitive or even on nervous structures, but on biophysical principles that are pervasive among biological organisms. In our work we explore this line of research by defining an emotion model that is inspired on basic mechanisms that have been proposed as sustaining the structure and activity of biological organisms. Underlying the proposed model is a view where "basic biological organization is brought about by a complex web of energy flows" [Popa 2004]. Since the proposed model is based on the interchange between an organism and its environment, expressed mainly as energy flows, we called it *flow model of emotion.* Its overall structure will be presented next.

2.1 The Flow Model of Emotion

Two main aspects are recognized as fundamental to emotional phenomena: (i) the relation between the agent and its environment - for instance Lazarus identifies the "person-environment relationship" as the "basic arena of analysis for the study of the emotion process" [Lazarus 1991]; (ii) the agent's ability to cope with the current situation [e.g. Leventhal and Scherer 1987]. Underlying these two aspects is another central concept, motivation, that is, the driving force that directs agent behavior [e.g. Berridge 2004].

In cognitive appraisal models, emotional states result from the assessment of the agent-environment relationship by deliberative mechanisms according to agent's goals [e.g. Gratch et al. 2006, Hudlica 2005]. On the other hand, in behavioral models, emotional states result from a tight agent-environment coupling in the context of stimulus-response patterns and homeostatic drives [e.g. Cañamero 2003, Velásquez 1998]. These two descriptive levels have however known limitations, in particular the difficulty to model the dynamic non-linear nature that characterizes emotional phenomena [Scherer 2000]. Some approaches have been proposed to address this problem [e.g. Breazeal 2002, Hudlica 2005], however they are mainly based on parameterized intensity and decay functions defined empirically. Is there another descriptive level that can provide an adequate support to address this problem? The answer to this question could be yes, if we consider that emotional phenomena may not be rooted on cognitive or even on nervous structures, but on more fundamental biophysical principles.

To address that level, an adequate model of agent organization and internal structures and mechanisms is needed. That model must be able to represent thermodynamic aspects that occur at a biophysical level. An adequate base support is the notion of *dissipative structure* [Nicolis and Prigogine 1977]. Dissipative structures are open systems governed by the interchange of energy with the environment and able to maintain themselves in a state far from equilibrium, yet keeping an internally stable overall structure. The maintenance of that internal stability in spite of environmental changes is done through feedback networks that motivate the system to act.

These feedback networks incorporate the basic notion of motivation [e.g. Berridge 2004] by maintaining in definite ranges of values internal and external variables that represent some form of energetic potential. The maintenance of a basic life support energy flow can be seen as a base motivation. From this base motivation other forms of motivation emerge according to the agent internal structure and organizational context, as is the case of *drives*, at an homeostatic level, or *desires*, at a deliberative level.

The basic flow of energy can be directly related to central aspects of emotional phenomena previously referred, namely, the agent-environment relationship and the relation between motivation and emotion. These two aspects are intrinsically associated in order to support the maintenance of the structure and activity of organisms through self-generation, a process known as *autopoieses* [Maturana and Varela 1987]. This is the basic context that motivates the flow model of emotion.

From the relation between an agent's internal motivation and its external situation, expressed through the energy flow that results from agent-environment interaction, behavioral dynamics arise that, from our point of view, constitute a particularly adequate support to characterize emotional phenomena in a way that is independent of the type or level of complexity of the agent.

From a thermodynamic point of view, to achieve its motivations an agent must apply an internal potential to be able to produce the adequate change in the environment. However, the concretization of the intended change depends also on the characteristics of the current environmental situation. That agent-environment relation can be modeled as a coupling conductance. Therefore, the process underlying motivation achievement can be modeled as a relation between an agent's internal potential, its *achievement potential*, and the agent-environment coupling conductance, the *achievement conductance*. The achievement potential represents the potential of change that the agent is able to produce in the environment to achieve the intended state-of-affairs. The achievement conductance represents the degree of the environment's conduciveness or resistance to that change, which can also mean the degree of environment change that is conducive, or not, to the agent intended state-of-affairs.

The achievement potential can be viewed as a force (P) and the achievement conductance as a transport property (C). The agent-environment relation underlying motivation achievement can therefore be characterized as a flow, called *achievement flow* (F), which results from the application of a potential P over a conductance C. From the relation between achievement potential and achievement conductance, expressed as achievement flow, internal dynamics arise that underlie agent behavior. These behavioral dynamics, expressed as energy flows, are considered the root of emotional phenomena, underlying and modulating cognitive activity. They are described as a vectorial function *ED*, called *emotional disposition* [see e.g. Morgado and Gaspar 2008]. The *emotional disposition* notion is defined as an action regulatory disposition or tendency, but it does not constitute in itself an emotion. In the proposed model, phenomena such as emotions and moods arise from the background effect of emotional dispositions across different organizational levels, generating increasingly reach emotional phenomena, especially at self-reflexive and social levels.

3 Modeling Emotional Agents

Although inspired by biophysical analogies, the main aim of the proposed model is to support the development and implementation of emotional agents, independently of their kind or level of complexity. Therefore it is necessary to concretize the base notions of the model in a computationally tractable way.

The first aspect that we need to address is the notion of energy. In thermodynamics, energy is usually defined as the capacity to produce work. In the context of the proposed model, energy can be defined as the capacity of an agent to act or, in a wide sense, to produce change. Considering an agent as a dissipative structure, that change is oriented towards the achievement of motivations, driven by internal potentials and expressed through energy flows.

For the implementation of software agents this thermodynamic level of description is however difficult to use directly due to its level of detail. For that reason we present a more abstract level of representation where these notions are characterized in a way that can act as a bridge between different levels of organization.

3.1 Internal Representational Structures

Energetic potentials can aggregate to form composite potentials. These composite potentials can represent different elements of the internal representational structures of an agent, such as a perception, a memory or an intention. Therefore they are broadly called *cognitive elements*. In this context we are using the term *cognitive* in the sense proposed by Maturana and Varela [1987], as a global property of an agent expressed through the ability of effective action in a given environment. This broader sense of cognition, distinct from the specific sense of symbolic information processing commonly associated to deliberative agent models [e.g. Newell 1990], allows to relate functionally similar concepts at different levels of organization, which is an important aspect of the proposed model.

Cognitive elements play different roles in agent internal activity. Three main roles can be identified: *observations*, *motivators*, and *mediators*. *Observations* result from perception processes, representing the current environmental situation. They can also result from simulated experience [Morgado and Gaspar 2005]. *Motivators* represent intended situations, acting as motivating forces driving agents' behavior. *Mediators* describe the media that support action, forming an interface between internal processing and action.

According to the organizational level, cognitive elements may have different kinds of instantiation. For example, in simple agents motivators may correspond to regulatory potentials, while in deliberative agents motivators may correspond to desires or high-level goals. In the same way, in simple agents mediators may correspond to direct mappings between perception and action, while in deliberative agents planning processes produce sequences of mediators that are translated by action processes into concrete action. In what concerns overall internal structure and processes, simple agents are composed of a fixed number of cognitive elements and very simple internal processes. Their behavior is directly guided by the dynamics resulting from the achievement potentials and flows [Morgado and Gaspar 2004], leading to basic adaptive behavior such as the *kineses* of some organisms [e.g. bacterial *chemotaxis*, Staddon 2001]. In more complex agents, internal representations are dynamically formed and changed, constituting an internal model [e.g. Newell 1990] based on which high-level cognitive processing, such as reasoning and decision-making, can occur.

3.2 Cognitive Space

To concretize the flow model of emotion, an approach to agent modeling was developed that follows a signal based metaphor in order to abstract the underlying thermodynamic characterization [Morgado and Gaspar 2007]. A main feature of that approach is the definition of a signal space [Oppenheim et al. 1997] that enables the description of agent's internal representational structures by means of geometrical concepts, therefore designated *cognitive space*. Conceptually, the cognitive space notion is related to a similar notion used in psychology to describe and categorize mental constructs based on a geometrical n-dimensional space [e.g. Newby 2001] and to the notion of conceptual space proposed by Gärdenfors (2000) based on a geometric treatment of concepts and knowledge representation. In particular, Gärdenfors proposes the notion of conceptual space as a way to overcome the opposition between the traditional, symbolic representations and the connectionist, sub-symbolic representations. This same motivation underlies the notion of cognitive space in our model. Relevant distinctions from these approaches exist however, since the proposed notion corresponds to a signal space where not only qualitative but also quantitative information can be described and interaction between the elements represented in the space is possible.

3.3 Modeling Emotional Dynamics

The motivation achievement process, from which emotional dispositions arise, can be described independently of the organizational level at which it occurs by a relation between a current situation, represented by an *observation*, and an intended situation, represented by a *motivator*. The internal activity of an agent is consequently guided by the maximization of the change (flow) that leads to the reduction of the distance between *observations* and *motivators* through the use of *mediators*. In the cognitive space this process can be described by the movement of cognitive elements, where motivators and observations correspond to specific positions and mediators define directions of movement, as illustrated in Fig. 8.1.

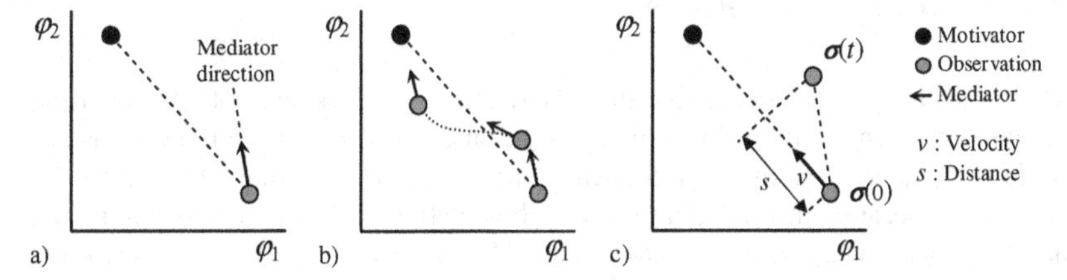

Fig. 8.1 Achievement of a motivator in a two-dimensional cognitive space (φ_1 and φ_2 represent the signals that define the space and $\sigma(t)$ represents a cognitive element along time).

As shown in Fig. 8.1.a, the direction of the selected mediators may not be the exact direction towards the motivator. Besides that, motivators can change and the dynamics of the environment (either internal or external) determines the movement of the observations. Fig. 8.1.b shows a possible trajectory resulting from the adjustment of agent's behavior by switching to different mediators. Independently of the specific processes that generated the new mediators, the forces that led to that change underlie the cognitive dynamics of an agent. From the point of view of the proposed model, emotional phenomena are considered the expression of those forces, characterized as emotional dispositions.

4 Adaptive Reasoning Mechanisms

The proposed model provides a generic framework for the implementation of agents of different types and levels of complexity. For instance, simple agents have a predetermined cognitive structure and very simple cognitive processes. Their behavior is directly guided by the dynamics resulting from the cognitive potentials and flows [Morgado and Gaspar 2004], leading to basic adaptive behavior such as the kineses of some organisms, e.g. bacterial chemotaxis [Staddon 2001]. However, the relation between emotional and cognitive phenomena becomes particularly relevant when we consider high-level cognitive processing, such as reasoning and decision-making, as experimental evidence indicates [Damásio 2000].

In the proposed model, the relation between emotional and cognitive phenomena is expressed in two ways: (i) by regulating the cognitive activity due to the present achievement conditions; (ii) by modulating the changes in the cognitive structure due to past experiences (i.e. the formation of emotional memories).

Next we will address the first aspect, that is, how do the emotion-based mechanisms support the adaptation of cognitive activity in order to achieve effective behavior under resource bounded conditions.

4.1 Focusing Mechanisms

The focusing mechanisms enable the adaptation of the type and rate of cognitive activity according to the achievement conditions, in order to make the best use of resources that are limited. We refer achievement conditions and not environmental conditions because this adaptation must be relative to what is relevant to the achievement of the agent motivators. This adaptation involves focusing cognitive processes along two perspectives: (i) a spatial perspective that refers to the space of cognitive elements over which processing can occur; and (ii) a temporal perspective that refers to the time available for cognitive processing. Focusing cognitive processes along both of these perspectives is a key issue in the implementation of autonomous agents able to handle the real-time requirements and resource bounded conditions typical of real-world domains. In the proposed model, these two perspectives of adaptation correspond to two main mechanisms, the attention focusing mechanism and the temporal focusing mechanism. Both of them depend on the signals produced by the emotional disposition mechanisms to operate. fig 8.2 illustrates how these different mechanisms are interconnected.

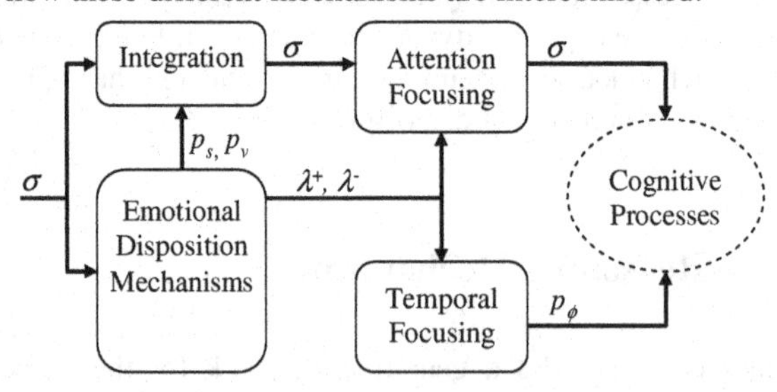

Fig. 8.2 Mechanisms underlying the adaptation of cognitive activity.

The affective signals λ^+ and λ^-, produced by the emotional disposition mechanisms, are directly input to both focusing mechanisms. On the other hand, the p_s and p_v cognitive potentials get integrated with the pair of cognitive elements that originated them, an observation and a motivator, constituting a composite cognitive element with an emotional disposition content. It is these composite cognitive elements that will be subject to attention focusing and possibly included in the attention field.

4.1.1 Attention Focusing

The attention focusing mechanism restricts the attention of the cognitive processes to specific cognitive elements according to their emotional disposition content (i.e. p_s and p_v cognitive potentials). This mechanism acts like a depletion barrier, producing an *attention field* formed by the cognitive elements able to bypass the barrier. Only the elements in the attention field are considered by the high-level cognitive processes, such as reasoning and deliberation.

The depletion barrier is characterized by a depletion intensity and by a permeability. The depletion intensity ε, is regulated by the affective signals λ^+ and λ^-, expressing their cumulative effect. That is:

$$\frac{d\varepsilon}{dt} = \alpha^+.\lambda^+ + \alpha^-.\lambda^- \quad (1)$$

where the sensitivity coefficients α^+ and α^- determine the influence of λ^+ and λ^- signals, respectively. In this way the intensity of the depletion barrier reflects the prevailing affective character and not the instantaneous values resulting from isolated experiences.

The permeability μ determines the intensity ε^σ of the interaction between a cognitive element σ and the depletion barrier, defined as:

$$\varepsilon^\sigma = \mu_s.p_s^\sigma + \mu_v.p_v^\sigma \quad (2)$$

where μ_s and μ_v are the permeability coefficients that determine the influence of p_s and p_v cognitive potentials. Given a certain depletion intensity ε, a cognitive element σ bypasses the barrier and is included in the attention field if $\varepsilon^\sigma > \varepsilon$.

4.1.2 Temporal Focusing

The temporal focusing mechanism regulates the rate of cognitive activity by providing a time-base for overall cognitive processing. This time-base corresponds to a signal p_ϕ with frequency ω_ϕ, which determines a period for cognitive processing. That is, it determines the time available before some behavior must be produced.

The regulation of the frequency ω_ϕ is determined by the affective signals λ^+ and λ^-. Like in the attention mechanism, sensitivity coefficients β^+ and β^- determine the influence of λ^+ and λ^- signals. In the same way, the frequency ω_ϕ expresses the cumulative effect of λ^+ and λ^- signals in order to reflect the prevailing affective character of the achievement conditions. That is:

$$\frac{d\alpha_\phi}{dt} = \beta^+.\lambda^+ + \beta^-.\lambda^- \tag{3}$$

By regulating the rate of cognitive activity, the temporal focusing mechanism has a direct influence on the balance between the use of computational resources and solution quality. For instance, it allows taking advantage of partial planning, anytime algorithms or other types of bounded reasoning mechanisms [e.g. Atkins et al. 2001, Dean et al. 1993]. Notice that the effect of the temporal focusing mechanism extends beyond what can be explored in this paper, since it can also have an influence on perception, action and memory formation.

5 Decision-Making Based on Long-Term Adaptation

In order to support real-time decision-making in uncertain and dynamic environments, long-term adaptation must be considered, i.e. the ability to take advantage of past experiences in order to adapt to the changing conditions and to anticipate future situations. In the proposed agent model that ability is related do autobiographic emotional memory, as will be presented next.

5.1 Emotional Memory

As cognitive elements change with time they describe trajectories in the cognitive space that reflect past experiences. These trajectories can be assimilated into the cognitive structure of an agent, forming autobiographic memories.

The cognitive elements that constitute those trajectories have an integrated emotional disposition, therefore they form *emotional disposition memories*, which can be related to what is referred by other authors as *emotional memories* [e.g. Damásio 2000, Bower 1994]. Cognitive potentials and cognitive elements were previously described as evolving in time. However time can be viewed as a special dimension in cognitive space. In this way, autobiographic memories are organized according to the time of formation, which is expressed as a specific spatial frequency. In the same way that the quality-representation frequencies ω allow for qualitative discrimination, a time-representation frequency ν allows for memory discrimination (that is, discrimination between different occurrence instants). Considering an abstract spatial dimension x, the memory mechanisms generate a time-reference signal $\phi_\nu(x)$ whose spatial frequency ν continuously and monotonically changes with time.

In agents with memory, this signal is used to modulate each cognitive element $\sigma(t)$ to produce a cognitive element $\sigma(t,x)$, which can be incorporated into the memory or just be used to interact with the memory. That is:

$$\sigma(t,x) = \sigma(t)\phi_v(x) \quad (4)$$

In the cognitive space this new representation of a cognitive element can be related to the previous one (4) as follows. Considering some instant $t = \tau$, a cognitive element $\sigma(t,x)$ is represented in a cognitive space CS_K as a vector $\boldsymbol{\sigma}$, defined as:

$$\boldsymbol{\sigma} = \rho_v \cdot (\rho_0, \rho_1, \ldots, \rho_k) \quad (5)$$

where the dimensional factor $\rho_v \in \mathbb{C}$ conveys the intensity and frequency v of signal $\phi_v(x)$ at time $t = \tau$. From this expression it is clear that time is not represented as just another qualitative dimension. Rather, time has an effect on each quality dimension representation.

Due to its signal-based nature, the memory field of the cognitive structure of an agent acts as a resonant structure where cognitive elements are activated by qualitative and temporal contact. Therefore, a trigger cognitive element $\boldsymbol{\sigma}$ interacting with the memory field will produce multiple recalled memory elements. Given a memory element $\boldsymbol{\sigma}_M$ previously assimilated, a recalled memory element $\boldsymbol{\sigma}_R$ is formed as follows:

$$\boldsymbol{\sigma}_R = \langle \boldsymbol{\sigma}, \boldsymbol{\sigma}_M \rangle \cdot \boldsymbol{\sigma}_M \quad (6)$$

where $\langle \boldsymbol{x}, \boldsymbol{y} \rangle$ denotes the scalar product of vectors $\boldsymbol{x}$ and $\boldsymbol{y}$. In this way, the recalled memory elements are modulated images of the original memory elements, whose intensity depends on the correlation (expressed by the scalar product) between the trigger element and the memory elements [Morgado and Gaspar 2005a].

5.2 Integrating Memory and Attention Mechanisms

Given the possibly large number of memory elements that can be recalled, the agent must decide on what to focus or decision-making in due time will not be possible. The attention focusing mechanism allows to deal with this problem by restricting the attention of the cognitive processes to specific cognitive elements, namely recalled memory elements, according to their emotional disposition content. This mechanism acts like a depletion barrier, producing an *attention field* formed by the cognitive elements able to bypass the barrier. Only the elements in the attention field are considered by the high-level cognitive processes, such as reasoning and deliberation. Fig. 8.3 illustrates the relation between the different mechanisms involved.

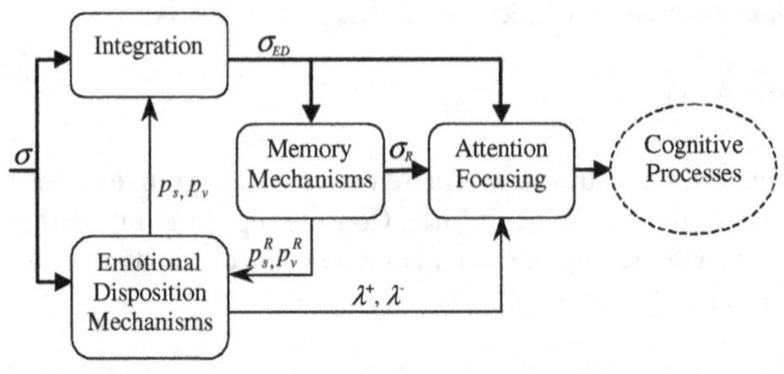

Fig. 8.3 Emotion-based mechanisms supporting adaptation and decision-making.

Besides producing the emotional disposition potentials p_s and p_v, the emotional disposition mechanisms also generate two regulatory signals λ^+ and λ^- that convey the affective character underlying those cognitive potentials (the intensity of positive and negative valences). Together with the emotional disposition potentials p_s and p_v, these signals are the main source for the regulation of cognitive activity, including the regulation of the attention field depletion barrier.

The depletion barrier is characterized by a depletion intensity and by a permeability. The depletion intensity ε, is regulated by the affective signals λ^+ and λ^-, expressing their cumulative effect. The permeability μ determines the intensity ε^σ of the interaction between a cognitive element $\boldsymbol{\sigma}$ and the depletion barrier. Given a certain depletion intensity ε, a cognitive element $\boldsymbol{\sigma}$ bypasses the barrier and is included in the attention field if $\varepsilon^\sigma > \varepsilon$.

At the same time, the emotional disposition content associated with the recalled memories (the p_s^R and p_v^R cognitive potentials shown in figure 3) is fed back to the emotional disposition mechanisms, producing an anticipatory regulatory effect, which can play a key role in highly dynamic environments where the time available for reasoning is limited.

6 Discussion

The relationship between emotion, reasoning and decision-making has been an active area of research [Hudlicka and Cañamero 2004], in particular after relevant theoretical and experimental work [e.g. Damásio 2000, LeDoux 2000] presented concrete evidence of this relationship. However, the focus of the work concerning emotions in intelligent agents has been on modeling and implementing emotional aspects and not much in combining emotion and cognition.

In this area, one main line of research has been on appraisal theory based models [e.g. Ortony et al. 1988]. Appraisal theories emphasize the structural aspects of emotion elicitation, but don't say much about the underlying processes [Smith and

Kirby 2001]. On the other hand, the aspects of adaptation and learning, which are directly related to emotion, as experimental studies show [LeDoux 2000] are not addressed.

A complementary line of research addresses these limitations by adopting a "design-based" approach [e.g. Sloman 2001] where emotional mechanisms are embedded within an overall architecture in a resource bounded agent. Although this approach explicitly addresses the problem of integrating emotion and cognition for intelligent behavior under resource bounded conditions, a sharp line is drawn between cognitive and emotional processing, where emotion plays essentially an interruptive role [Oatley and Johnson-Laird 2002, Simon 1967].

Our proposal departs from these approaches by modeling emotion and cognition as two symbiotically integrated aspects of agent cognitive activity. This means that the relation between emotion and cognition occurs not only at a functional specialization level. Instead it is intrinsic to all cognitive activity and to the nature of the involved cognitive elements. Recent experimental results support this view, indicating that in humans, emotion and higher cognition can be truly integrated, that is, at some point of processing, functional specialization is lost and emotional and cognitive influences inseparable [Gray et al. 2002].

On the other hand, two important aspects characterize emotional phenomena, the relation with adaptive behavior and the relation with reasoning and decision-making. This two-sided relation has remained almost unexplored in cognitive models due to the strong emphasis on functional division, which hinders the intrinsic relation of emotion and cognition as a whole. However, as some authors have proposed (e.g. Cañamero 2003), it is a fundamental aspect that enables effective intelligent behavior in concrete environments. This relation between adaptive behavior, emotion and high level cognitive processes is a main characteristic of the proposed model, enabling the adaptation to uncertain and dynamic environments and allowing the control of the high level cognitive processes.

References

Arzi-Gonczarowski Z (2002) AI Emotions: Will One Know Them When One Sees Them? Proc. 16th European Meeting on Cybernetics and Systems Research

Atkins E, Abdelzaher T, Shin K, Durfee E (2001) Planning and Resource Allocation for Hard Real-time, Fault-Tolerant Plan Execution. Autonomous Agents and Multi-Agent Systems, 4:57-78

Berridge K (2004) Motivation concepts in behavioral neuroscience. Physiology & Behavior, 81(2)

Bower G (1994) Some Relations Between Emotions and Memory. The Nature of Emotion: Fundamental Questions, eds. Ekman P, Davidson R, 303-305, Oxford Press

Breazeal C (2002) Designing Sociable Machines. MIT Press

Cañamero L (2003) Designing Emotions for Activity Selection in Autonomous Agents. Emotions in Humans and Artifacts, eds. Trappl R, Petta P, Payr S, MIT Press, 115-148

Carver C, Scheier M (2002) Control Processes and Self-organization as Complementary Principles Underlying Behavior. Personality and Social Psychology Review, 6

Damásio A (2000) A Second Chance for Emotion. Cognitive Neuroscience of Emotion, eds. R. Lane, L. Nadel, 12-23, Oxford Univ. Press

Dean T, Kaelbling L, Kirman J, Nicholson A (1993) Planning with Deadlines in Stochastic Domains. Proc. 11th National Conference on Artificial Intelligence

Frijda N (1986) The Emotions, Cambridge University Press

Gärdenfors P (2000) Conceptual Spaces: The Geometry of Thought. MIT Press

Gratch J, Marsella S, Mao W (2006) Towards a validated model of "emotional intelligence". Proc. 21st National Conference on Artificial Intelligence

Gray J, Braver T, Raichle M (2002) Integration of Emotion and Cognition in the Lateral Prefrontal Cortex. Proc. of the National Academy of Sciences USA, 99:4115-4120

Hebb D, Thompson W (1979) The Social Significance of Animal Studies. Handbook of Social Psychology, 2nd ed., eds. Lindzey G and Aronson E, Addison-Wesley

Hudlica E (2005) Modeling interactions between metacognition and emotion in a cognitive architecture. Proc. AAAI Spring Symposium, Technical Report SS-05-04, AAAI Press

Hudlicka E, Cañamero L (2004) Architectures for Modeling Emotion: Cross-Disciplinary Foundations. AAAI Spring Symposium, Technical Report SS-04-02

Lazarus R (1991) Emotion and adaptation. Oxford Press

LeDoux J (2000) Cognitive-Emotional Interactions: Listen to the Brain. Cognitive Neuroscience of Emotion, eds. R. Lane, L. Nadel, 129-155, Oxford Univ. Press

Leventhal H, Scherer K (1987) The relationship of emotion to cognition: A functional approach to a semantic controversy. Cognition and Emotion, 1

Maturana H,Varela F (1987) The Tree of Knowledge: The Biological Roots of Human Understanding. Shambhala Publications

Moore S, Oaksford M (2002) Emotional Cognition, John Benjamins Press

Morgado L, Gaspar G (2003) Emotion in Intelligent Virtual Agents: The Flow Model of Emotion. Intelligent Virtual Agents, eds. Rist T, et al., LNAI 2792, 31-38, Springer-Verlag

Morgado L, Gaspar G (2004) A Generic Agent Model Allowing a Continuous Characterization of Emotion. Proc. 17th European Meeting on Cybernetics and Systems Research

Morgado L, Gaspar G (2005) Emotion Based Adaptive Reasoning for Resource Bounded Agents. Proc. 4th International Joint Conference on Autonomous Agents and Multi-Agent Systems

Morgado L, Gaspar G (2005a) Adaptation and Decision-Making Driven by Emotional Memories. Progress in Artificial Intelligence, eds. Bento C, et al., LNAI 3808, Springer-Verlag

Morgado L, Gaspar G (2007) A Signal Based Approach to Artificial Agent Modeling. Proc. 9th European Conference on Artificial Life, LNCS 4648, Springer

Morgado L, Gaspar G (2008) Towards Background Emotion Modeling for Embodied Virtual Agents. Proc. 7th International Joint Conference on Autonomous Agents and Multiagent Systems

Newby G (2001) Cognitive space and information space. Journal of the American Society for Information, Science and Technology, 52(12)

Newell A (1990) Unified Theories of Cognition. Harvard Press

Nicolis G, Prigogine I (1977) Self-Organization in Nonequilibrium Systems: From Dissipative Structures to Order trough Fluctuations. John Wiley & Sons

Oatley K, Johnson-Laird P (2002) Emotion and Reasoning to Consistency. Emotional Cognition, eds. Moore S, Oaksford M, John Benjamins

Oppenheim A, Willsky A, Nawab S (1997) Signals & Systems. Prentice Hall

Ortony A, Clore G, Collins A (1988) The Cognitive Structure of Emotions. Cambridge Univ. Press

Popa R (2004) Between necessity and probability: searching for the definition and origin of life. Springer

Rolls E (2001) The brain and emotion', Oxford Press

Scherer K (2000) Emotions as Episodes of Subsystem Synchronization Driven by Nonlinear Appraisal Processes. Emotion, Development, and Self-Organization, eds. Lewis M, Granic I, 70-99, Cambridge Univ. Press

Simon H (1967) Motivational and Emotional Controls of Cognition. Psychological Review, 74:29-39

Sloman A (2001) Beyond Shallow Models of Emotion. Cognitive Processing, 2(1):178- 198

Smith C, Kirby L (2001) Affect and Cognitive Appraisal Processes. Affect and Social Cognition, ed. Forgas J, 75-92, L. Erlbaum

Staddon J (2001) Adaptive Dynamics: The Theoretical Analysis of Behavior. MIT Press

Velásquez J (1998) Modeling Emotion-Based Decision-Making. Proc. AAAI Fall Symposium Emotional and Intelligent: The Tangled Knot of Cognition, 164-169, AAAI Press

A Generic Recommendation System based on Inference and Combination of OWL-DL Ontologies

Hélio Martins and Nuno Silva

GECAD - Knowledge Engineering and Decision Support Research Group

School of Engineering – Polytechnic of Porto, Porto, Portugal

(email: {hamm, nps}@isep.ipp.pt)

Abstract. This paper proposes a generic information recommender system based on (i) OWL DL ontologies, (ii) the inference capabilities of generic classifiers and (iii) information provided from multiple sources. The system dynamically and automatically recommends facts from the user's repository according to the user's context, which is automatically inferred according to the information provided from multiple sensors (e.g. user's agenda, GPS, social networks, etc.). The system operates iteratively, such its output re-feeds the information repositories, and thereafter constraining the next iteration.

Keywords: Context-aware, Recommendation, Ontologies.

1 Introduction

Recommender systems support users by identifying interesting products and services in situations where the number and complexity of offers outstrips the user's capability to survey them and reach a decision [Felfernig et al. 2007]. Useful information for use during the recommendation process is the user context [Lilien et al. 1992]. Context is any information that can be used to characterize the situation (e.g. location, time/date, agenda, social relationships) of an entity. An entity is a person, place or object that is considered relevant to the interaction between a user and an application, including the user and applications themselves (Dey 2001). Usually, in many context-aware systems, context information is provided by sensors placed in the environment of the system, providing data that is interpreted/combined in order to identify the current context.

Ontologies promise to play a pivotal role for different semantic-based and semantic-aware applications [Luther et al. 2005]. For instance, the ongoing research and standardization efforts in the area of the Semantic Web [Berners-Lee et al. 2001] are focusing on a machine readable, meaningful description of the elements

A. Madureira et al. (eds.), *Computational Intelligence for Engineering Systems: Emergent Applications*, Intelligent Systems, Control and Automation: Science and Engineering 46,
DOI 10.1007/978-94-007-0093-2_9,

in the World Wide Web. In this paper we demonstrate how the semantic web recommended language OWL DL and reasoning inference engines can be applied to create a generic information recommender system.

Consider the following scenario: The end user knows many people and each one has many contact methods (e.g. work contact, friend contact, relative contact, mobile, fax, etc.). The main idea is to automatically recommend the most appropriate contact method to the end user according to contact's current context. To solve this problem, the generic process presented in Fig. 9.1 can be applied.

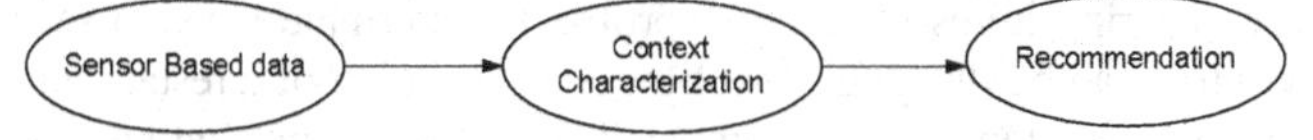

Fig. 9.1 Generic process to context-aware recommendation systems

- Sensor-based data - responsible for receiving the information from the sensors, and transforming it according to the system representation;
- Context characterization - responsible for characterizing the current context of the user according to the information provided by the previous process;
- Recommendation process - responsible for recommending information according the current context of the user.

Through a worked example we illustrate the logical foundations of OWL and show how basic reasoning support can be used to derive new knowledge and use it in a recommendation process. The worked example intends to recommend contact information from the user's contact list according to its current context, but this generic process is not for solving this specific example only but can be applied to a wide range of context-aware recommendation scenarios.

The rest of the paper is organized as follow: Section 2 discusses the background of the work, section 3 presents the proposed solution and the Section 4 presents conclusions and future work.

2 Background

2.1 Recommendation System

There has been a growth in interest in Recommender Systems in the last two decades [Adomavicius and Tuzhilin 2005], since the appearance of the first papers on this subject in the mid-1990s [Resnick et al. 1994]. The aim of such systems is to help users to find items that they might appreciate from huge catalogues. In this field, collaborative filtering approaches can be distinguished from content-based ones. The former is based on a set of user ratings on items, while the latter uses item content descriptions and user thematic profiles. In collaborative filtering, users can be compared based upon their shared appreciation of items, creating the notion of user neighborhoods. Similarly, items can be compared based upon the shared appreciation of users, rendering the notion of item neighborhoods. The item rating for a given user can then be predicted based on the ratings given in her user neighborhood and the item neighborhood. Content-based recommendation systems recommend an item to a user based upon a description of the item and a profile of the user's interests. Content-based recommendation systems share in common a means for describing the items that may be recommended, a means for creating a profile of the user that describes the types of items the user likes, and a means of comparing items to the user profile to determine what to recommend. Item descriptors can be the genre of a film or the location of a restaurant, depending upon the type of item being recommended. Finally, items that have a high degree of proximity to a given user's preferences would be recommended. While collaborative filtering systems often result in better predictive performance, content-based filtering offers solutions to the limitations of collaborative filtering, as well as a natural way to interact with users. These complementary approaches thus motivate the design of hybrid systems.

2.2 OWL language and reasoning

The OWL (Web Ontology Language) is a language for defining and instantiating Web ontologies. An OWL ontology may include descriptions of classes, properties and their instances. Given such an ontology, the OWL formal semantics specifies how to derive its logical consequences, i.e. facts not literally present in the ontology, but implied by the semantics [W3C\OWL 2004]. OWL DL supports those users who want the maximum expressiveness without losing computational completeness (all deductions are guaranteed to be computed) and decidability (all computations will finish in finite time) of reasoning systems. OWL DL is so named due to its correspondence with description logics [DL], a field of research that has studied a particular decidable fragment of first order logic. OWL DL was designed to support the existing Description Logic technological infrastructure and has desirable computational properties for reasoning systems.

Logically well-founded ontologies not only offer ways for describing a domain of interest, but also allow reasoning about the represented information. While in practice the term "ontology" is sometimes overused and often specified imprecisely, we feel that the classical definition of a DL knowledge base best reflects its intended meaning [Baader et al. 2003]: in DL terminology, a tuple consisting of a T-box and an A-box is referred to as a knowledge base (the actual ontology). The T-box contains the intentional knowledge in terms of a terminology (hence the term "T-box") and is build through declarations that describe general properties of concepts. On the other hand the A-box contains extensional knowledge – also called assertional knowledge (hence the term "A-box") – that is specific to the individuals of the domain of discourse.

2.3 *Reference architecture*

In [Baltrunas 2008], the authors suggest generic system architecture for recommendation systems based on contextual information. This architecture includes the following process components (Fig. 9.2):

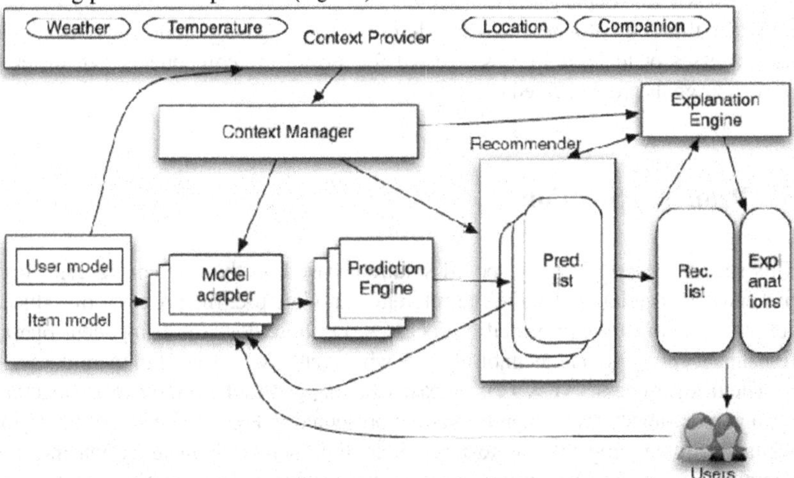

Fig. 9.2 Reference architecture for context-aware recommendation system (Baltrunas 2008)

- Context provider – is a module that tracks changes of contextual variables. It contacts appropriate context services and stores all context states in a local database;
- Context manager – performs all the reasoning related to the context. It determines if a contextual variable is important for a prediction, removes noisy context data and makes predictions for missing contextual variables;
- Recommender – takes all the produced recommendation lists and combines them into a final recommendation list. It can use information obtained from the contextual manager to filter out, or change the ratings for the final list;
- Explanation engine – takes the final list of recommendations and provides the explanations for each of them. It closely collaborates with the recommendation engine to identify the best recommendations. In fact, a recommendation may be suggested just because it could be easy to explain and not only because it has a high predicted rating. The explanation engine could also use the contextual reasoner to find out the necessary information to motivate a recommendation because of the particular contextual conditions.

Furthermore, the reference architecture includes the following data components:

- The User and Item – represent user and item information in the system. The user is modeled with his preferences and in Collaborative Filtering this is a

vector of item ratings. The item model captures relevant knowledge in the application domain;
- Prediction list – list of predictions generated by the prediction engine;
- Explanations – list of explanations generated by the explanation engine;
- Recommendation – list of recommendations produced by the recommender with possible explanations;

While this is an all-encompassing description, it is also generic and abstract, as no processes or technology is suggested for any of the components, but instead, only their goals are described.

3 Proposed Solution

This section describes the proposed solution, based on the reference architecture presented in Section 2. Unlike the reference architecture, the proposal introduces several components with specific processes (i.e. how does the component operationalies its process) and technology (i.e. what technology is used in the process).

The three processes (**Fig. 9.1**) can easily be mapped to the reference architecture for a context-aware recommender system presented in Fig. 9.2: the sensor based information integration into context provider, the context characterization into the context manager, the recommendation into the context. In the following sections we will explain each of these steps.

3.1 Sensor-based data

By definition, a sensor is a device that measures a physical quantity and converts it into a signal which can be read by an observer or by an instrument [Wikipedia\Sensor 2009]. In our approach a sensor is an abstract entity that captures relevant information from other systems or devices to the system.

The goal of this component is to transform data provided by multiple sources into the system representation. For that, usually 3 steps are followed [Martins 2007]:

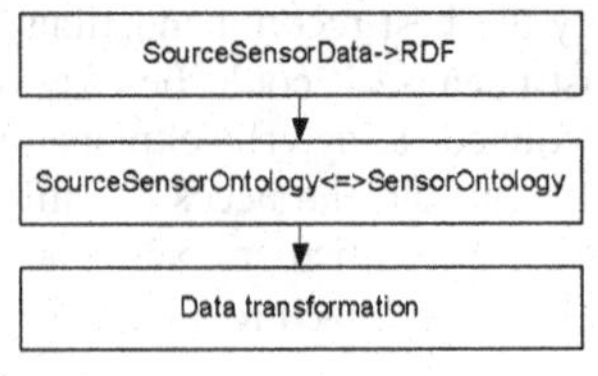

Fig. 9.3 Sensor data transformation steps

- SourceSensorData->RDF – In this step the data from a sensor (e.g. GPS) is converted to RDF format according to a *SourceSensorOntology* (e.g. GPS ontology)
- SourceSensorOntology<=>SensorOntology – In this step a map between the *SourceSensorOntology* and *SensorOntology* (Ontology used in the system) is created resulting in a mapping document.
- DataTransformation – In this step the sensorData based on the SourceSensorOntology is transformed into data based on the *SensorOntology* through the mapping document.

According to our running example, the following sensors may be useful to provide information to the system:

- GPS – can provide information about the current user location;
- Social Network – can provide information about people that the user interacts with. There are many kind of social network (family, professional, friendship) that help identify what kind of relation exists between the participants;
- Agenda – Can provide information about what the user is doing.

All this information provided by these sensors can be helpful to the recommendation. However there are many aspects that should be considered as constraints:

- Data heterogeneity, with respect to syntax, structure and semantics, leading to different granularity, periodicity, relevance, inconsistency, etc.
- Similar data for different sources – different sensors may provide the same data, leading to redundancy, contradiction and ambiguity.

Because of these aspects, the data provided by the sensors should be preprocessed in order to reduce the data to common semantics and representation. In order to lift, normalize and integrate data from multiple sources, we apply the MAFRA Toolkit [Silva 2004] alignment and integration features. Further details are out of the scope of this paper.

Further, our proposal is to use semantic web technologies such as ontologies expressed in OWL DL and generic reasoning engines, to help solve such problems.

The sensory data is modeled through an ontology that defines the classes of data. Ontology starts defining the foundational classes of data in the form of primitive classes (i.e. those constrained by necessary conditions only). Location, DateAndTime and Appointment are examples of such classes. Yet, based on their properties, the ontology further specializes these classes into definable classes (i.e. those constrained by necessary and sufficient conditions), such as WorkMeeting (i.e. an Appointment whose subject is “Work”) or DinnerTime (i.e. occurring between “19h00” and “21h00”).

3.2 *Context categorization*

User context categorization is based on the information provided by the sensors and a process for matching the observed context with the set of identified contexts.

For that, an ontology of context categories is necessary. The context ontology includes a set of classes representing the identified contexts. Different classes and sub-classes of contexts are categorized and distinguished through the detailed descriptions of their particular properties restrictions, whose values will be provided by the sensors.

User's contexts are identified and specified in ontologies expressed in OWL DL, exploiting the necessary and sufficient conditions modeling feature. The necessary and sufficient conditions are defined according to relations to sensor-acquired integrated data, such as Location (arriving from GPS) and Appointment-Type (arriving from the user's agenda). This ontology modeling approach provides a way of specifying the context classes based on the sensory data and the automatic inference-based classification of context. Fig. 9.4 illustrates a simple example of a defined context class.

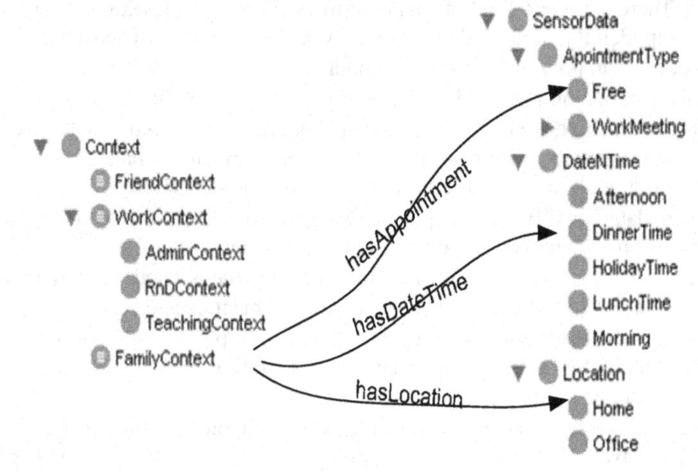

Fig. 9.4 Context ontology example

In this example, the user context would be classified as a FamilyContext if the sensory data confirms that (i) the user's current Location is "Home", (ii) the user's current AppointmentType is "Free" and (iii) Time is "DinnerTime". These are necessary and sufficient conditions as suggested by the icon before FamilyContext (≡).

The inference process is then able to categorize the single instance of context to the most specific sub-class of context, depending on the sensory data stored in properties' values. Yet, because contexts are not necessarily disjoint, sensory data might lead to different acceptable categorizations of the context. Also, sometimes the sensory data does not allow inferring of a more specific type, in which case the context is not further categorized. Otherwise, the context classifications will constrain the recommendation process.

3.3 Recommendation

Traditionally, recommender systems deal with applications that have two types of entities: users and items. The same consumer may use different decision-making strategies and prefer different products or brands, under different contexts [Bettman et al. 1991]. According to [Lilien et al. 1992], "consumers vary in their decision-making rules because of the usage situation, the use of the good or service (for family, for gift, for self) and purchase situation (catalog sale, in-store shelf selection, sales person aided purchase)." Therefore accurate prediction of consumer preference undoubtedly depend on the degree to which we have incorporated the relevant contextual information in a recommender system.

Furthermore, this paper claims that the user's context drastically influences the information he/she prefers, thus playing a major role in the recommendation process and in the outcome. Again, the proposed approach includes an ontology that classifies the information to recommend and the application of generic inference mechanism to categorize it.

The domain ontology defines the domain concepts according to other domain concepts and property restrictions (e.g. Person has Contact). In fact, considering that a specific class of domain information is the main focus of the recommendation, many other complementary domain concepts might exist in order to characterize it. In the running example, while the main recommendation concept is the Person, the examples of complementary information would be Contact class and Appointment class. The complementary domain concepts are instantiated from different data sources, such as agenda, social networks and (free-text) documents repositories. Nevertheless, the domain concepts are always classified according to the context type, i.e. one of the constraints of the domain concept is the type of Context.

Accordingly, the worked example is further developed by defining an ontology for categorization of Person. **Fig. 9.5** presents the Worker concept. The conceptual interpretation of this definition states that a Worker is a Person who "has" a ContactWork "in" a WorkContext (i.e. the context is classified as WorkContext). For example, a Person is classified as a worker not only because he/she has a work contact but also because the context is about work (i.e. WorkContext).

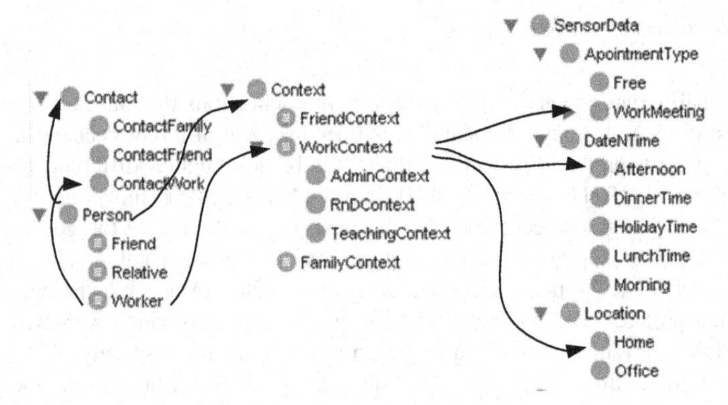

Fig. 9.5 Domain ontology, context ontology and sensor data ontology combination

3.4 Two-step generic recommendation

A significant problem of recommender systems is that they are domain dependent (e.g. movies [Plantié et al. 2005], books [Lindenet al. 2003]). In dynamic environments, the domain of recommendation is always changing (e.g. I want movie recommendations on Friday night, book recommendations on Monday morning). As we explain in previous sections, the system is generic enough to be applied to any ontology or domain that satisfies the imposed development conditions (e.g. three distinct hierarchies, necessary and sufficient conditions).
However, to be applied to any domain, the system needs firstly to decide which domain it will recommend information for. Thus, we propose a system with two steps of recommendation as illustrated in Fig. 9.6.

In **Step 1** the system decides which domains are relevant in current contexts. The *Recommendation 1* process will recommend the domain ontology(ies). Therefore, this recommendation is similar to the one explained in section 3.3, but because the system is intended to recommend domain ontologies, instead of using any ontology, only the ontology of ontologies will be used. For that, the following data and ontologies are employed:

- Context ontology used to describe contexts;
- Inferred contexts according to the context ontology and sensorData ontology (see section 3.2);
- Ontology of ontologies used to describe domain ontologies, i.e., describes witch ontologies can be recommended to be the domain ontologies.

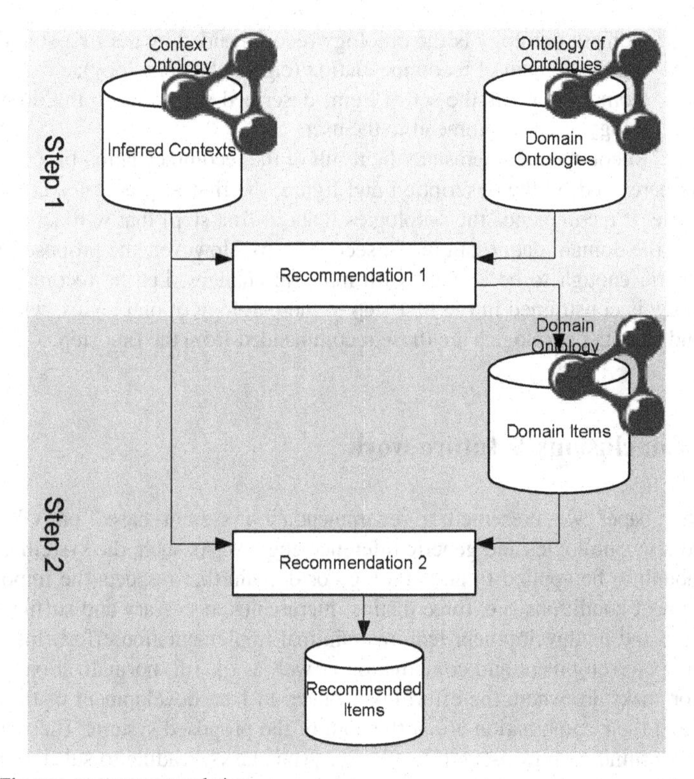

Fig. 9.6 The two step recommendation

Fig. 9.7 presents an example of the ontology of ontologies where there are types of domain ontology like contacts (our running example) or movie that can be used in the recommendation process (recommendation 2);

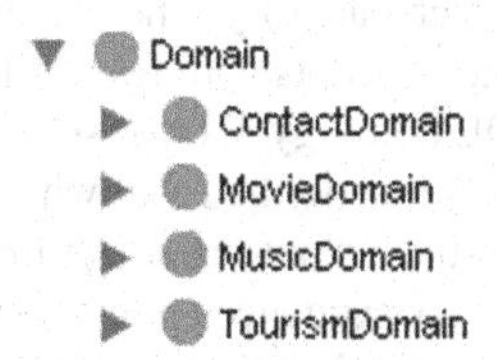

Fig. 9.7 Ontology of ontologies example

- Domain Ontologies is a set of domain ontologies described according to the ontologies of ontologies.

The result of the recommendation made in **step 1** (domain ontology) is used in **step 2** where the recommendation of domain information is recommended. The following components are employed:

- Domain ontology is the ontology recommended by the first step describing the domain of recommendation (e.g. contact ontology);
- Domain items is the set of items described according to the domain ontology to be recommend to the user;
- Recommended items are the result of the recommendation process.

As perceived by the description and figure, the first step actuates at the meta-level, i.e. it recommends the ontologies (data in first step) that will act as ontologies of the domain data (data in the second step). However, the proposed process is generic enough to be applied with minimal changes, i.e. the recommendation ontology is constrained in the first step to the ontology of ontologies, while in the second step, the ontologies are those recommended from the first step.

4 Conclusions & future work

In this paper we presented a recommendation system based on OWL DL-expressive ontologies and generic inference engines. As such, the system is generic enough to be applied to any ontology or domain that respects the imposed development conditions (i.e. three distinct hierarchies, necessary and sufficient conditions) and its development requires minimal implementation effort. Instead, the ontology development and combination as well as the lift, normalization and integration tasks are where the effort takes place. In fact, development of the ontologies and their combination are at the core of the proposed system. These are very time-consuming, human-centric and error-prone tasks, leading to substantial difficulties in setting up and maintaining the system. Moreover, in order to carry out these tasks two competencies are required but seldom observed in a single person: ontology engineering and domain expertise. This issue has been observed in the application of the proposed approach to a new domain. In this experience, two librarians were asked to categorize documents in a hierarchy, defining the necessary and sufficient conditions for each category. The results were disappointing as they were unable to systematize the concepts with respect to their properties in order to represent them in an OWL DL ontology. Conversely, we asked two proficient ontology engineers to model the same domain, but while they were able to represent their conceptualization, the result did not match the librarians' conceptualization.

In order to minimize these consequences, the research efforts are now directed to the automatic generation and combination of the ontologies based on the definition of simple primitive classes (e.g. Person, Contact) and factorization points. Based on the envisaged ontology factorization process, the definable classes are automatically generated according to the combination dimensions defined by the user, i.e. two primitive hierarchies and a combination vector. For example, if the Worker class has to be factorized with the WorkContext, it would give rise to a hierarchy in which Worker is specialized in WorkerInAdminContext, WorkerInRnDContext and WorkerInTeachingContext, such necessary and sufficient conditions are automatically established (Fig. 9.8).

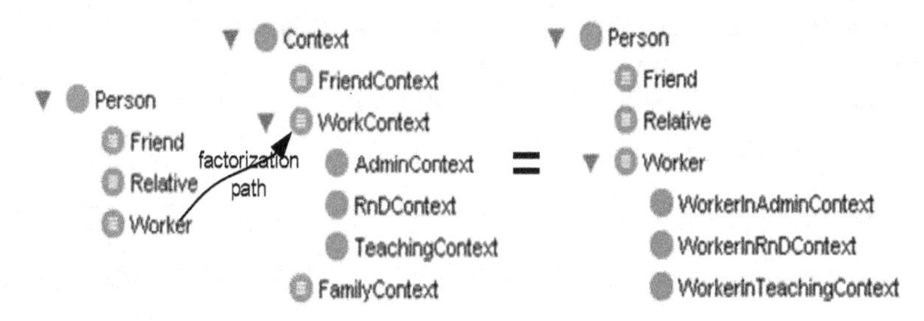

Fig. 9.8 Ontology factorization example

The usefulness and applicability of this process is an open issue and needs to be further investigated. It is likely that the process will generate huge unmanageable ontologies.

References

Adomavicius G, Tuzhilin A (2005) Toward the next generation of recommender systems: A survey of the state-of-the-art and possible extensions. in Knowledge and Data Engineering, IEEE Transactions on., 17(6):734-749

Alexander F, Friedrich G, Schmidt-Thieme L (2007) Guest Editors' Introduction: Recommender Systems. in IEEE Intelligent Systems, 18-21. CA,USA

Baader F, Calvanese D, McGuinness D, Nardi D, Patel-Schneider P (2003) "Theory, Implementation, and Applications." In Description Logic Handbook. Cambridge University Press

Baltrunas L (2008) Exploiting contextual information in recommender systems. In RecSys '08: Proceedings of the 2008 ACM conference on Recommender systems, 295--298. Lausanne, Switzerland

Berners-Lee T, Hendler J, Lassila O(2001) The semantic web: Scientific American

Bettman, J. R., E. J., Johnson, and Payne J. W. (1991) Consumer Decision Making. In Handbook of Consumer Behavior, 50-84

Dey K (2001) "Understanding and Using Context." In Personal and Ubiquitous Computing

Lilien G, Kotler P, Moorthy S (1992) Marketing Models. Prentice Hall

Linden G, Smith B, York J (2003) Amazon.com recommendations: item-to-item collaborative filtering. in Internet Computing, IEEE, 76--80

Luther M, Mrohs B, Wagner M, Steglich S, Kellerer W (2005) Situational reasoning - a practical OWL use case. In Autonomous Decentralized Systems, 461--468

Martins, H (2007) "Suporte a conversão e introdução de conteúdos baseado em ontologias

Plantié M, Montmain J, Dray G (2005) Movies Recommenders Systems: Automation of the Information and Evaluation Phases in a Multi-criteria Decision-Making Process. In Database and Expert Systems Applications. Springer Berlin / Heidelberg

Protégé (2009) http://protege.stanford.edu/

Resnick P, Iacovou N, Suchak M, Bergstorm P, Riedl J (1994) GroupLens: An Open Architecture for Collaborative Filtering of Netnews. In Proceedings of ACM 1994 Conference on Computer Supported Cooperative Work

Silva N (2004) Multi-Dimensional Service-Oriented Ontology Mapping
W3C\OWL (2004) http://www.w3.org/2004/OWL/
Wikipedia\Sensor (2009) http://en.wikipedia.org/wiki/Sensor

GIGADESSEA – Group Idea Generation, Argumentation, and Decision Support considering Social and Emotional Aspects

Goreti Marreiros[1,2], Ricardo Santos[1,3] and Carlos Ramos[1,2]

[1]GECAD – Knowledge Engineering and Decision Support Group

Porto, Portugal,{goreti,csr}@dei.isep.ipp.pt

[2]Institute of Engineering – Polytechnic of Porto, Porto, Portugal

[3]School of Management and Technology of Felgueiras – Polytechnic of Porto

Felgueiras, Portugal, rjs@estgf.ipp.pt

Abstract. Today we are faced with a Global Economy that creates many difficulties for joining groups at the same space. New trends appeared, like Distributed and Ubiquitous meetings. In this paper we present a model to an Emotional-aware Group Support System covering Idea Generation, Argumentation, and Decision Making processes.

1 Introduction

There is a clear advantage involving groups of persons in the most important processes of Management, like in Idea Generation, Argumentation and Decision Making. By discussing and combining ideas, counter-ideas, critical opinions, identifying constraints, and alternatives a group of individuals is able to test better possible solutions for identified problems, sometimes in the form of new products, services, and plans.

On the last decades, Operations Research, Artificial Intelligence, and Computer Science have achieved a tremendous success in creating software systems able to achieve optimal solutions, even for complex problems. The only drawback is that the human being does not agree always with these solutions. Sometimes the problem is related with the incorrect parameterization of the problem. But more frequently, the human being does not like the solution due to aspects that cannot be quantified, most of times due to affective aspects like emotion, mood, and personality.

Monolithic individual Decision Support Systems centred on optimizing solutions are being substituted by Collaborative systems and Group Decision Support Systems more centred in establishing the connection between people in organiza-

A. Madureira et al. (eds.), *Computational Intelligence for Engineering Systems: Emergent Applications*, Intelligent Systems, Control and Automation: Science and Engineering 46,
DOI 10.1007/978-94-007-0093-2_10,

tions. These systems follow a kind of Social paradigm. Combining both approaches (optimization-centred and social-centred) is something that is being experimented. However, even if we achieve a support system with the ability to combine in the right way the optimization with the collaborative nature of the work, we will still miss a very important point: the emotional nature of groups' participants in tasks involving idea generation, argumentation, and decision making.

In paper we intends to address some of the points referred above. This GIGADESSEA (Group Idea Generation, Argumentation, and Decision Support considering Social and Emotional Aspects) project envisages the development of Support Systems for Idea Generation, Argumentation, and Decision Making, involving groups and considering the emotional aspects in the process.

2 Background

In this section we will perform a brief description of the main areas covered by GIGADESSEA project: idea generation, argumentation, group decision, and emotion.

2.1 Idea Generation

Idea Generation and creativity are topics studied since the 60's [McGrath 1984] and still to the present day new studies are published [Buzan 2006]. This fact can be explained because organizations and groups wish to improve the effectiveness and efficiency of their thoughts and interactions, so, several studies in the field of psychology and group interaction analysis have been the catalyst of the area [McGrath 1984, Yakemovic and Conklin 1990].

Known techniques are brainstorming [McGrath 1984], brainwriting [Paulus and Yang 2000], goal oriented idea generation [Oshiro et al 2003], nominal groups [Lago et al. 2007], and IBIS [Yakemovic and Conklin 1990]. Later, computers and technology have revived the matter allowing the analysis and comparison of existing methods with computer assisted variants [Adrianson and Hjelmquist 1991]. New techniques for generating ideas [Oshiro et al 2003] appeared, including knowledge representation and semantics. Several techniques have proved that are able to improve the performance of the meetings. With the support provided it is intended to maximize important process gains like more information, synergy, learning, stimulation, more objective evaluation and the minimization of process losses like domination of some members, failure to remember among others [Nunamaker et al. 1991]. Some commercial well known system are Big Think (http://www.bigthink.com/), Xmind (http://www.xmind.com), and GroupSystems [Nunamaker et al. 1991].

2.2 Argumentation

In the recent view of Multi-Agent Systems, Argumentation is referred as a key form of interaction among autonomous agents [Parson et al. 1998,Kraus et al. 1998]. Inside the context of multi-agent negotiation, an argument is seen as a piece of information able to influence other agents negotiation stance and/or justify the own agent negotiation stance [Jennings et al. 1998].

Katia Sycara developed the Persuader system, a framework for intelligent conflict resolution and mediation [Sycara 1990]. This work led to subsequent research by Kraus [Kraus et al. 1998], which proposed a logic model for reaching agreements through argumentation. In this model the argumentation style is based on the psychology of persuasion, using threats, rewards and appeals.

In our previous work, we have been developing work in argumentation field. We have applied argumentation techniques in group decision context, where argumentation is used either by agents or humans to justify positions or to persuade group members [Marreiros et al. 2009].

2.3 Group Decision Making

Group decision-making represents an important role in actual organizations. The new economy demands that the decisions must be taken quickly and without the degradation of the quality of the decision-making process or its results.

According to Huber [Huber 2001] a Group Decision Support System (GDSS) consists of a set of software, hardware, languages components and procedures that support a group of people engaged in a decision related meeting. A more recent definition says that GDSS are interactive computer-based environments which support concerted and coordinated team effort towards completion of joint tasks [Nunamaker 1997].

Generically we may say that GDSS aims to reduce the loss associated to group work (e.g. time consuming, high costs, improper use of group dynamics, etc.) and to maintain or improve the gains (e.g. groups are better to understand problems and in flaw detection, participants' different knowledge and processing skills allow results that could not be achieved individually). The use of GDSS allows groups to integrate the knowledge of all members into better decision making.

Modelling groups and/or social systems is a hard task. During the last years several GDSS have been developed aiming supporting groups where members are dispersed in time and space, and applied to several areas [Marreiros et al. 2007,Ramos et al. 2009,Dennis and Garfiels 2003,Adkins et al. 2003].

2.4 Emotion

The terms emotion, mood and affect are many times used alternately. According to Forgas [Forgas 1995] affect is the most generic and usually is used to refer to mood and emotion. Emotion is normally referred as an intense experience, of short duration (second to minutes), with a specific origin and in general the individual is conscious of that. In contrast, moods have a propensity to be less intensive, longer lasting (hours or even days) and remain unconscious for the individual. Moods may be caused by an intense or recurrent emotion, or yet by environmental aspects. In Psychology literature several examples could be found on how emotions and moods affects the individual decision making process

[Forgas 1995,Loewenstein and Lerner 2003,Schwarz 2000,Barsade 2002]. For instance, individuals are more predisposed to recall memories that are congruent with their present emotional state. There are also experiences that relate the influence of emotional state in information seeking strategies and decision procedures.

The emotional state of an individual varies along the time and influences his behaviour and interactions with other group members.

The process of emotional contagion is the tendency to express and feel emotions that are similar to and influenced by those of others. This process could be analyzed based on the emotions that a group member is feeling or based on the group members mood

[Neuman and Strack 2000].

Ortony, Clore and Collins developed the OCC model which [Ortony et al. 1988] is widely used for emotion simulation of embodied agents [Gratch and Marsella 2006,Bída and Brom 2008], namely inside our group [Marreiros et al. 2008]. In OCC model, agents concerns in an environment are divided into goals, standards, and preferences and twenty-two emotion labels are defined. Due to the complexity in his original model Ortony proposed a new simplified model with 10 emotional categories divided in five positive categories (joy, hope, relief, pride, gratitude and love) and five negative categories (distress, fear, disappointment, remorse, anger and hate)[Ortony 2003].

3 Proposed Model

In fact it is quite common that when meetings are fundamental we find important people for these meetings in different points of the globe, contacting clients, suppliers, government institutions, etc. Thus, one aspect that needs to be covered is this distributed and ubiquitous nature of the groups. But one question need to be answered: "Is the interaction the same, emotionally speaking, when we have a face-to-face meeting or a long-distance participation?"

Bellow we will present a model to support an Emotional-aware Group Decision process covering Idea Generation, Argumentation, and Decision Making tasks.

3.1 Model

GIGADESSEA follows the McGraph tasks model for group work. McGrath [1984McGrath 1984] built a model joining different kinds of tasks performed by a group, based on the analysis of the work group, the carried tasks and the procedures associated to the tasks. Two different axis and four quadrants compose it. The axis describes tasks and attitudes necessary to decision making execution. Conceptual or behavioural distinction is made by the horizontal axis and the vertical axis categorizing the tasks in order to support conflict resolutions or collaboration attitudes.

GIGADESSEA tools support the decision process in 3 of the four McGrath quadrants. They are: the generation, choose and the negotiation. The execution quadrant was excluded because here are performed tasks on behavioural domain and that required only the execution of what was decided and documented.

GIGADESSEA is composed by 4 main modules: IdeaGeneration, WebMeeting, a relational database system and pervasive hardware in a meeting room (Fig. re 10.1).

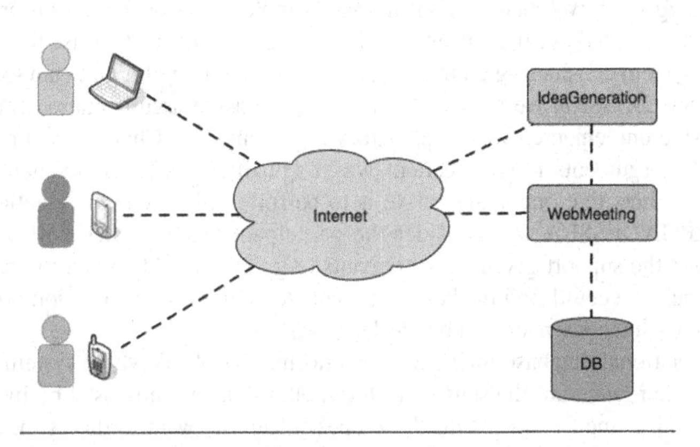

Fig. 10.1 GIGADESSEA architecture

The main blocks of the system are: IdeaGeneration and WebMeeting that will described in the following sections.

a. Idea Generation

The Idea Generation module will support a ubiquitous group decision meeting dedicated to the idea generation task. Emotional aspects will be considered in order to identify how these aspects can contribute for the improvement of important process gains like for instance more information, synergy, learning, stimulation, more objective evaluation and the minimization of process losses like domination of some members, failure to remember and production blocking.

In our previous work we already developed a prototype of an Idea Generation system, IGTAI [Ramos et al. 2009]. This system was conceived to be included into an Intelligent Decision Room, LAID [Marreiros Et Al. 2008]. However IGTAI did not consider the emotional aspects, but it can be a good starting point for the inclusion of these aspects in Idea Generation process.

b. WebMeeting

WebMeeting is a module that aims helping geographically distributed people and organisations in solving multi-criteria decision problems namely supporting the selection of alternatives, argumentation, voting techniques and meeting setup.

In our previous work we used persuasive argumentation [Parson et al. 1998] to model the argumentation process (generation, selection and evaluation) that supports the group decision making [Marreiros et al. 2008,Marreiros Et Al. 2008]. In this project we develop an agent based system that simulates group decision processes considering emotional and argumentative aspects. We made a set of experiences using the developed system in some simple real-world decision problems. The results pointed out the advantage of meeting participants that use the decision support system considering emotional aspects in the phase of argumentation.

In GIGADESSEA we intend also to use persuasive argumentation [Parson et al. 1998], complemented with explanatory argumentation. Choice based on an exchange of arguments is an excellent way to justify possible preferences and to convince others that one's alternative is to be followed. The argumentation process in GIGADESSEA will consider the participant profile which will be fundamental for the support given to participants (e.g. if we could predict the profile of our partners we could find the best arguments to be used in the decision process in order to reach a consensus or a better decision).

The relational database maintains the information of the whole system and the pervasive hardware are the sensors of the toolkit that are composed by interactive screens and a smart board. With these ones we are allowed to detect interactions between the meeting members minimizing the effort of users.

3.2 Scenario

As test scenario we are proposing the area of tourism because of the constant necessity to gather new tourists. In this area there is exists always the need to de-fine new products, like touristic packages. So, the proposal of a new Touristic Package can be seen as a new idea, which needs to be argued and voted.

Suppose this simple scenario, a group of friends intends to choose a place to stay during the weekend, some prefers to go to the beach, others to the mountains, others to a SPA and others to an European city to visit museums. This divergence of preferences may cause a conflict between them. One way to solve the problem and reach a consensual decision (that could be one of the referred above or other) is using to argumentation to justify choices or to convince others about a specific proposal. During this decision process that may include idea generation tasks, several emotions and affective states are triggered updating the emotional stat, which will allow to evaluate the context-aware emotional model.

So in a first phase we will use the example referred above as test case and in a second phase we intend to test the system by offering to companies in the market to use our GIGADESSEA system in order to discuss new tourism packages.

GIGADESSEA system will not particularly conceive for Tourism area. GIGADESSEA should be seen as having any other possible application domain; the only constraint is that we need to find some experts of that domain available to test GIGADESSEA system.

4 Conclusions

The work proposed in GIGADESSEA is a complete innovation. Several systems have been implemented with more or less success in the areas of Idea Generation support or Group Decision Making. Few systems have been implemented for Argumentation in a Group perspective. What is completely new is the inclusion of Emotional aspects in Groups activities and the integration of the three phases of the process: Idea Generation, Argumentation, and Decision Making.

The main results of the project will be the implementation and test of an Emotional-aware Group Support System covering Idea Generation, Argumentation, and Decision Making (GIGADESSEA system), and its use in a real scenario for defining new Touristic packages.

Our main goal with GIGADESSEA will is to contribute for the creation of the futuristic concept of Intelligent Decision Rooms [Marreiros Et Al. 2008,Ramos et al. 2009], seen as smart spaces that support the human being in the most valuable aspects of his activity.

References

Adkins M, Burgoon M, Nunamaker JF (2003) Using group support systems for strategic planning with the United States Air Force. Decision Support Systems, 34 (3), pp. 315-337.

Adrianson L and Hjelmquist E (1991). Group processes in face-to-face and computer-mediated communication, Behaviour & Information Technology, vol. 10, pp. 281-296

Barsade S (2002). The Ripple Effect: Emotional Contagion and Its Influence on Group Behavior. Administrative Science Quarterly, 47, pp. 644-675

Bída M and Brom C (2008). Towards a platform for the education in emotion modeling based on virtual environments. In: Proceedings of the 3rd Workshop on Emotion and Computing: Current Research and Future Impact. (Ed. Dirk Reichardt) ISSN 1865-6374, Kaiserslautern, Germany, September 23rd, 2008. 45 - 52

Buzan T (2006) Mind Map Book, Pearson Education

Dennis AR, Garfield MJ (2003) The adoption and use of GSS in project teams: Toward more participative processes and outcomes. MIS Quarterly: Management Information Systems, 2 (2), pp. 289-323

Forgas J (1995) Mood and judgment: The affect infusion model (AIM). Psychological Bulletin,117, pp. 39-66

Gratch J and Marsella S (2006). Evaluating a computational model of emotion. Journal of Autonomous Agents and Multiagent Systems, 11(1), pp. 23-43

Huber, GP (1982) Group decision support systems as aids in the use of structured group management techniques. Proc. of second international conference on decision support systems, San Francisco in C. W. Holsapple, A. B. Whinston, Decision support systems: a knowledge-based approach (Thomson Learning, inc, 2001)

Jennings NR, Parsons S, Noriega P and Sierra C (1998) On Argumentation-Based Negotiation. Proc. Int. Workshop on Multi-Agent Systems, Boston, USA

Kraus S, Sycara K, Evenchik A (1998) Reaching agreements through argumentation: a logical model and implementation. Artificial Intelligence, vol.104 n.1 p.1-69

Lago P, Beruvides M, Jian J, Canto A, Sandoval A, Taraban R (2007) Structuring group decision making in a web-based environment by using the nominal group technique. Computers & Industrial Engineering, vol. 52, pp. 277-295

Loewenstein G and Lerner JS (2003) The role of affect in decision making, in Handbook of Affective Sciences. Davidson, R., Scherer, K., Goldsmith, H., eds., Oxford University Press

Marreiros G, Machado J, Ramos C and Neves J (2009). Argumentation-based Decision Making in Ambient Intelligence Environments. International Journal of Reasoning-based Intelligent Systems (IJRIS)

Marreiros G, Ramos C and Neves J (2007) Modelling group decision meeting participants with an Agent-based approach. International Journal of Engineering Intelligent Systems, Vol. 15, No.3. ; CRL

Marreiros G, Santos R, Freitas C, Ramos C, Neves J and Bulas-Cruz J (2008) LAID - a smart decision room with ambient intelligence for group decision making and argumentation support considering emotional aspects, International Journal of Smart Home, vol. 2, n. 2, pp.77-94

Marreiros G, Santos R, Ramos C, Neves J and Bulas-Cruz J (2008) ABS4GD: A Multi-agent System that Simulates Group Decision Processes Considering Emotional and Argumentative Aspects. AAAI Spring Symposium Series March 26-28, Stanford University

Marreiros G, Santos R, Ramos C. (2009) Smart Offices and Intelligent Decision Rooms, in Handbook of Ambient Intelligence and Smart Environments(AISE), H. Nakashima, J. Augusto, H. Aghajan (ed.), Springer

McGrath J (1984(Groups: Interaction and Performance; Prentice-Hall, Englewood Cliffs, N.J.

Neumann R and Strack F. (2000). Mood contagion: The automatic transfer of mood between persons. Journal of Personality and Social Psychology, 79 pp 211-223

Nunamaker JF, Dennis A, Valacich J, Vogel D, George J (1991). Electronic meeting systems", Communications of the ACM, v.34 n.7, p.40-61

Nunamaker, JF et al. (1997) Lessons from a dozen years of group support systems research: A discussion of lab and field findings. Journal of Management Information Systems, Vol. 13 No.3

Ortony A (2003) On making believable emotional agents believable, In R. P. Trapple, P. (Ed.), Emotions in humans and artefacts, Cambridge: MIT Press

Ortony A, Clore GL Collins A. (1988). The cognitive structure of emotions, Cambridge: Cambridge University Press

Oshiro K, Watahiki K and M. Saeki (2003) Goal-Oriented Idea Generation Method for Requirements Elicitation. Proceedings of the 11th IEEE International Conference on Requirements Engineering, IEEE Computer Society, p. 363

Parsons S, Sierra C, Jennings N (1998). Agents that reason and negotiate by arguing. Journal of Logic and Computation, Vol. 8 No. 3 pp. 261-292.

Paulus P and Yang H (2000) Idea Generation in Groups: A Basis for Creativity in Organizations. Organizational Behavior and Human Decision Processes, vol. 82, , pp. 76-87

Schwarz N (2000) Emotion, cognition, and decision making. Cognition and Emotion, 14(4), pp. 433-440

Sycara K (1990). Persuasive Argumentation in Negotiation. Theory and Decision, Vol. 28,No. 3, pp. 203-242

Yakemovic, K and Conklin, E.J (1990) Report on a development project use of an issue-based information system. Proceedings of the 1990 ACM conference on Computer-supported cooperative work, Los Angeles, California, United States: ACM, pp. 105-118

Electricity Markets: Transmission Prices Methods

Judite Ferreira, Zita Vale and Hugo Morais

GECAD – Knowledge Engineering and Decision-Support Research Group of the Institute of Engineering – School of Engineering

Polytechnic of Porto (ISEP/IPP), Rua Dr. António Bernardino de Almeida, 4200-072 Porto, Portugal

(e-mail: {mju, zav, hgvm}@isep.ipp.pt)

Abstract In the context of electricity markets, transmission pricing is an important tool to achieve an efficient operation of the electricity system. The electricity market is influenced by several factors; however the transmission network management is one of the most important aspects, because the network is a natural monopoly. The transmission tariffs can help to regulate the market, for this reason transmission tariffs must follow strict criteria.
This paper presents the following methods to tariff the use of transmission networks by electricity market players: Post-Stamp Method; MW-Mile Method Distribution Factors Methods; Tracing Methodology; Bialek's Tracing Method and Locational Marginal Price.
A nine bus transmission network is used to illustrate the application of the tariff methods.

1 Introduction

Traditionally the power sector was based on vertically integrated utilities covering generation, transmission, distribution, and retail. In the last years this organization has evolved towards vertical disintegration. In the new organization the generators compete in the wholesale power market to sell their energy to the buyers (distributors, retailers and consumers). Under this new unbundled structure of generation and transmission with competitive trading between wholesale market participants, the traditional pool operating functions have been more clearly defined and segregated into Market operation functions and System operation functions. The market operation functions are associated with energy trading, scheduling, and settlement of energy transactions in different time horizons and the system operation functions are related to operation and control of the bulk power system to meet load and security needs. It must assure a real time dispatch to balance supply and de-

A. Madureira et al. (eds.), *Computational Intelligence for Engineering Systems: Emergent Applications*, Intelligent Systems, Control and Automation: Science and Engineering 46,
DOI 10.1007/978-94-007-0093-2_11,

mand, manage ancillary services, maintain system reliability, and manage transmission congestion [Rothwell and Gómez 2003].

An effective competition in reformed wholesale electricity markets can only be achieved if the following five prerequisites are met: separation of the grid from generation and supply; wholesale price deregulation; sufficient transmission capacity for a competitive market and non-discriminating grid access; excess generation capacity developed by a large number of competing generators; an equilibrium relationship between short-term spot markets and the long-term financial instruments that marketers use to manage market price volatility; an essentially hands-off government policy that encompasses reduced oversight and privatization. The absence of any one of the first five conditions may result in an oligopoly or monopoly market whose economic performance does not meet the efficiency standards of a competently managed regulated electrical utility [Dicorato et al 2003].

An important issue to guarantee effective wholesale electricity competition concerns the division of responsibilities into owning, operating, and regulating the transmission system. A cornerstone of restructuring is the separation of generation from transmission. Under this segregated structure, transmission companies and systems operators continue to be regulated. All Transmission Systems operators (TRANSCOs) must provide comparable and nondiscriminatory service to independent generators. The availability of transmission installations and the location of new transmission investments significantly affect trading opportunities.

In spite of the deregulated electricity market to have separated the activities of the transmission and production, the electrical transmission networks are assumed as a natural monopoly. This is due to the economic (and sometimes even physical) impossibility of the existence of several alternative infrastructures as transmission networks. Anyway, this monopoly cannot constitute an obstacle for the activities of the agents who act in these markets. So, adequate regulation mechanisms that guarantee fair and non discriminatory access to the transmission electrical network are required [Tan et al 2008]. Many authors argue that the transmission network is a key element in the process of liberalizing the electricity market. The transmission tariff system and the user costs allocation must preserve an adequate resource allocation among market agents [Abhyankar et al 2006].

Transmission tariff methods can be evaluated analyzing how adequately they reflect the real impact of network users in the transmission network.

2 Methodologies for Transmission Cost Allocation

All transmission tariff methodologies present advantages and disadvantages. Each methodology using different ways to allocate the costs due to the use of the transmission network.

The Post-Stamp and the contract path are two primarily methods for transmission allocation [Abhyankar et al 2006].These methods are simple and can be easily implemented, but they do not consider the actual transmission network use and do not provide adequate economic signals aiming at efficient transmission network use. Some methods, as the MW-mile method [Madiji et al 2008], consider the power flow impact that each agent causes in each line of the transmission network. Marginal methods are other methodology for allocation operational as well as embedded costs of transmission networks.

Different methodologies for transmission cost allocation have appeared in recent years. The *pro-rata* method allocates costs to generators and loads according to the sum of active power produced/consumed by each generator/load.

Other methods that are more complex distribute the cost based on the active power flow produced by generators and demands through the transmission lines. These methods use the proportional sharing principle; the flow attributed to each generator and load in upstream lines, determines the power flows through downstream lines. Thus, these flows are associated with the origins and destinations, i.e., generators and loads. More details of this method can be found in [Bialek 2007] and [Kirschen et al 1997].

The nodal methods are used in many countries to allocate the network usage costs- This allocation is based on optimal power flow sensitivity in each line due to the power injected at each bus.

The network usage method uses equivalent bilateral exchanges to allocate costs to generators and loads. Thus, each load is attributed a generation fraction and a fraction of each load is attributed to each generator. The cost attribution by the network usage method occurs considering the impact in terms of power flow of each equivalent bilateral exchange in each transmission line obtained by DC power flow calculation.

The Z_{bus} method [Conejo et al 2007] presents a solution based on the Z_{bus} matrix and considers the current injection at each bus. The combination of Z_{bus} matrix and currents injections determines the measure of sensitivity that indicates what the individual contribution of each current injection to produce the power flow through of a transmission line.

2.1 Post-Stamp Method

The Postage Stamp method is traditionally used by electric utilities to allocate the fixed transmission cost among the users firm transmission service [Abhyankar et al 2006]. This method is an embedded cost method, also called rolled-in embedded method. It does not require power flow calculations and is independent of the transmission distance and network configuration. In other words, the charges associated with the use of the transmission system determined by the postage stamp method are independent of the transmission distance, supply and delivery points or the loading on the different transmission facilities caused by the transactions under

study. The method is based on the assumption that the entire transmission system is used, regardless of the actual facilities that carry the energy of the actual energy transactions that are taking place. The method allocates charges to a transmission user based on an average embedded cost and magnitude of the user's transacted power.
Expressions (1) e (2) allow to evaluate the tariff $R(u)$ imputed to each transaction u.

$$P = \frac{CT}{\sum_g PG(g)} \tag{1}$$

$$R(u) = P \times W(u) \tag{2}$$

where

CT is the total cost to share ($)
$W(u)$ is the active power transitioned (MW)
PG(g) is total active power production (MW)

The Post-Stamp method has several disadvantages because does not give economical signals. This method to tariff the transmission only reflects the energy transitioned independently of the way each transaction impact on the transmission network.

2.2 MW-Mile Method

The MW-Mile method allocates the costs of the transmission system proportionally to the MW power flows caused by a transaction in a transmission network [Madiji et al 2008]. This method is an embedded cost method that is also known as a line by line method because it considers, in its calculations, changes in MW transmission flows and transmission line lengths. The tariff P is evaluated using expression (3):

$$P = \frac{CT}{\sum_k F_k \times L_k} \tag{3}$$

where

CT is the total cost to share ($)
F_k is the power flow in line k (MW)
L_k is the line length (mile)

This method evaluates charges associated with each wheeling transaction based on the transmission capacity use as a function of the transacted power and of the path followed by the transaction. This method depends on the actual operational system configuration usually. Usually it uses dc power flow calculation but it can also use ac power flow calculation.

2.3 Base Method

Although the Base Method is usually considered to be a variant of the MW-mile method, there is a great difference between this method and the original MW-Mile [Abhyankar et al 2006]. While the original method considers the maximum capacity of the line in denominator, this method uses the total flow of the line. With the Base Method the total cost of the system is distributed by all the transactions, according to expression (4) that allows to calculate R(u).

$$R(u) = \sum_k C_k \frac{F_k(u)}{\sum_s F_k(s)} \tag{4}$$

With this method, some of the R(u) taxes can be negative, what means that one definitive transaction can receive a credit to use the transmission system. This happens when the active power flow provoked by a transaction is in the opposite direction to the one of the active power flow in the initial conditions of the system. Effectively, this is important when a line works near to its maximum capacity.

2.4 Module or Use

The Module or Use method distributes the total cost of the system among the several transactions, considering transactions in both directions. In this way, all transactions are due to pay transmission use tariffs, but the cost is more distributed. The expression (5) shows how to determine the tax to be paid by the R (u) transaction [Tan et al 2008].

$$R(u) = \sum_k C_k \frac{|F_k(u)|}{\sum_s |F_k(s)|} \tag{5}$$

The difference of this method when compared with the Base method is that it considers the absolute values of the line flows originated by each transaction *u*, instead of the signed values.

2.5 Zero Counterflow

The Zero Counterflow method only taxes the positive flows. This method assumes that the negative flows are beneficial for the network. Therefore, in these cases, the transactions are not paid but they do not generate a credit. The expression (6) shows how to determine the tax to be paid by the R(u) transaction [Tan et al 2008].

$$R(u) = \begin{cases} \sum_k C_k \dfrac{F_k(u)}{\sum_s FD_k(s)} & \text{for } F_k(u) > 0 \\ 0 & \text{for } F_k(u) \leq 0 \end{cases} \tag{6}$$

$$FD_k(u) = \begin{cases} F_k(u) & \text{if } F_k(u) > 0 \\ 0 & \text{if } F_k(u) \leq 0 \end{cases} \tag{7}$$

The function $FD_k(u)$ only considers the impact, provoked by the transaction u in line k, when transaction u increases the active power flow in this line.

2.6 Dominant Flow

The Dominant Flow method merges the concepts that are on the basis of the two previous methods (Module or Use and Zero Counterflow). The Dominant Flow method considers that R(u) is the sum of two terms RA(u) and RB(u). The tem RA(u) is determined using the Zero Counterflow method substituting cost C_k by CA_k. The term RB(u) is determined using the "Module or Use" method where C_k is substituted by CB_k. Factor CA_k corresponds to the cost due to the transit in the line for the base case of the system and CB_k corresponds to the cost of the non used capacity. The expression (8) shows how to determine the tax to be paid by the R(u) transaction [Tan et al 2008].

$$R(u) = RA(u) + RB(u) \tag{8}$$

where

$$\begin{cases} RA(u) = \begin{cases} \sum_k CA_k \dfrac{F_k(u)}{\sum_s FD_k(s)} & \text{For } F_k(u) > 0 \\ 0 & \text{For } F_k(u) \leq 0 \end{cases} \\ RB(u) = \sum_k CB_k \dfrac{|F_k(u)|}{\sum_s |F_k(s)|} \end{cases} \tag{9}$$

$$\begin{cases} CA_k = C_k \dfrac{FM_k - F_k(u)}{FM_k} \\ CB_k = C_k \dfrac{F_k(u)}{FM_k} \end{cases} \quad (10)$$

With this method all the participants that use the system in the opposite direction of the resultant flow receive an incentive, which consists of lower costs. This incentive increases when the system is more loaded, arriving to zero cost when the system is at the maximum load. These economical signs are coherent with the intents of reducing expansion costs.

2.7 Distribution Factors Methods

Distributions factors are sensitivity factors that allow evaluating the line flows. These factors are calculated based on linear load flows; they can be used to determine the impact of each generator and load on transmission line flows [Galetovic and Palma-Behnke 2008].

2.7.1 Generalized Generation Distribution Factors

The Generalized Generation Distribution Factors (frequently referred to as GGDFs or D factors) determine the impact of each generator in line flows. The expression used to calculate the GGDF is:

$$D_{ik,p} = D_{ik,r} + A_{ik,p} \quad (11)$$

The $A_{ik,p}$ factor is evaluated using the reactance matrix and dc power flow approximation. This factor represents the incremental use of the transmission network and depends only on the transmission network characteristics.

The GGDF factor of the reference bus can be obtained as follows:

$$D_{i\,k,r} = \frac{P_{i\,k}^{0} - \sum_{\substack{p=1 \\ p \neq r}}^{N} A_{i\,k,p} \cdot G_p}{\sum_{p=1}^{N} G_p} \quad (12)$$

The impact on line flows evaluated by this method is not an incremental value but an absolute value. GGDF factor depends on the lines characteristics and it also depends on the reference bus.

2.7.2 Generalized Load Distribution Factors

The Generalized Load Distribution Factors (GLDFs or C factors) determine the impact of each load on line flows.

$$C_{i\,k,m} = C_{i\,k,r} + A_{i\,k,m} \tag{13}$$

The GLDF factor of the reference bus can be obtained as follows:

$$C_{i\,k,r} = \frac{P_{i\,k}^{0} - \sum_{\substack{m=1 \\ m \neq r}}^{N} A_{i\,k,m} \cdot L_m}{\sum_{m=1}^{N} L_m} \tag{14}$$

The impact on line flow evaluated by this method is not an incremental value but an absolute one. GLDF factor depends on the lines characteristics, but it does depend of the reference bus.

2.8 Tracing Methodology and Bialek's Tracing Method

This section explains the principles concepts of the tracing methodology and the proportional sharing. The second part of this section show how to apply this math technique in the Bialek's tracing algorithm.

2.8.1 Tracing Methodology

The core of the power flow tracing method [Galetovic and Palma-Behnke 2008] is the power flow proportional sharing principle ahich says that, for every node in a network, the proportion of power flow on each outflow branch fed by each inflow branch is equal to the proportion of the inflow from this branch in the total in-flows.

For example, in figure 11.1, there are 4 branches connected with node i, where q_j and q_k are the inflow branches and q_m and q_l are the outflow branches.

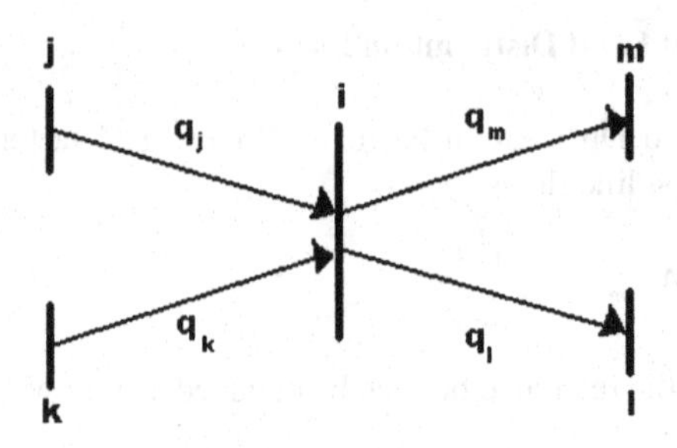

Fig.11.1 Proportional Sharing [Galetovic and Palma-Behnke 2008]

Taking into account the Kirchhoff laws, we have:

$$q_j + q_k = q_m + q_l \tag{15}$$

Let us assume that nodes j and k can represent producers connected directly to bus i, and node m and node l represent consumers connected to node i.
The outflow q_m is composed by two components in function of q_k and q_j, as in expression (16):

$$q_m = \frac{q_j}{q_j + q_k} * q_m + \frac{q_k}{q_j + q_k} * q_m \tag{16}$$

q_l, q_j, q_k can be evaluated in a similar way:

$$q_l = \frac{q_j}{q_j + q_k} * q_l + \frac{q_k}{q_j + q_k} * q_l \tag{17}$$

$$q_j = \frac{q_m}{q_m + q_l} * q_j + \frac{q_l}{q_m + q_l} * q_j \tag{18}$$

$$q_k = \frac{q_m}{q_m + q_l} * q_k + \frac{q_l}{q_m + q_l} * q_k \tag{19}$$

This way to evaluate the shares can be applied to all nodes of the transmission network. The goal is to calculate the power flow impact provoked by the transactions.

2.8.2 Bialek's Tracing Methodology

Tracing methods are based on ac power flow methods aiming to determine the contribution of transmission users to transmission network usage. Tracing methods can be used for transmission pricing and recovering fixed transmission costs. In this work, we discuss one tracing method, usually known as the Bialek's tracing method. This method assumes that nodal inflows are shared proportionally among nodal outflows [Galetovic and Palma-Behnke 2008, Zhaoxia J and Fushuan 2005] and it is used to evaluate the contribution of the generators for the losses in each transmission network branch.

To show how this method is applied, let us define the gross demand as the sum of a particular generator and its allocated part of the total transmission loss. The total gross in a system is equal to total actual generation. Topological distribution factors are given by expression 20 which $D^{g}_{ik,p}$ refers to the kth generator's contribution to line *i-k* flow.

$$p^{g}_{ik} = \frac{p^{g}_{ij}}{p^{g}_{i}} \sum_{p=1}^{n} \left[A_u^{-1}\right]_{ip} P_{G_p} = \sum_{p=1}^{n} D^{g}_{ij,p} P_{Gp};\ j \in \alpha^{d}_{i} \tag{20}$$

Where

$$p^{g}_{ik} = \sum_{k \in \alpha^{u}_{i}} \left|P^{g}_{ik}\right| + P_{G_i}; \qquad i = 1, 2, \ldots, n \tag{21}$$

$$\lfloor A_u \rfloor_{ik} = \begin{cases} 1 & i = k \\ -\dfrac{\left|P^{g}_{ik}\right|}{P_k} & k \in \alpha^{u}_{i} \\ 0 & otherwise \end{cases} \tag{22}$$

The gross power value at any bus is equal to the generated power amount at the bus taking into account the imported power flows from neighboring bus. The total usage of the network by the kth generator U_{Go} is calculated by summing up the individual contributions (multiplied by lines weights) of this generator to line flows. U_{Gp} value is given by:

$$U_{Gp} = \sum_{i=1}^{n} \sum_{k \in \alpha^{d}_{i}} W^{p}_{ik} D^{p}_{ik,p} P_{G_p} = P_{G_p} \sum_{i=1}^{n} \left\{ \frac{\left[A_u^{-1}\right]_{ip}}{p^{p}_{i}} \sum_{j \in \alpha^{d}_{i}} C_{ik} \right\} \tag{23}$$

2.9 Locational Marginal Price

In competitive electricity markets, Locational Marginal Prices (LMP) are important pricing signals for the participants as the effects of transmission losses and biding constraints are embedded in LMP values. While these LMP provide valuable information at each location, they not provide a detailed description in terms of contribution terms. The LMP components, on the other hand, show the explicit decomposition of LMP into contribution components, and thus, can be considered as better market signals [Sood et al 2007].

After solving congestion and to evaluate the losses, the standard locational price for location *i* and time *t* is calculated as:

$$LMP_i = LMP^{energy} + LMP_i^{loss} + LMP_i^{cong} \tag{24}$$

where

LMP_i	locational marginal price at bus i (\$/MWh)
LMP^{energy}	marginal energy price of system (\$/MWh)
LMP_i^{loss}	marginal loss price at bus i (\$/MWh)
LMP_i^{cong}	marginal congestion price at bus i (\$/MWh).

The loss and congestion components are defined as follows:

$$LMP_i^{loss} = (DF_i - 1) * LMP^{energy} \tag{25}$$

$$LMP_i^{cong} = -\sum_{l \in K} GSDF_l * \beta_l \tag{26}$$

where

DF_i	delivery factor at bus i
$GSDF_l$	generation Shift Factor at line l
B_l	constraint incremental cost (shadow price) associated with line k
k	set of congested transmission lines.

The shadow price is the change in the objective value of the optimal solution of an optimization problem obtained by relaxing the constraint by one unit. In a business application, a shadow price is the maximum price that management is willing to pay for an extra unit of a given limited resource.

Locational marginal pricing is a market-pricing approach used to manage the efficient use of the transmission system when congestion occurs on the bulk power grid.

2.9.1 Penalty Factors and Delivery Factors

The Penalty Factor associated with any bus on the transmission system is defined as the increase required in injection at that bus to supply an increase in withdrawn at the system reference bus with all other bus net injections held constant. Mathematically, the Penalty Factor for bus i can be calculated as:

$$PF_i = \frac{1}{\left(1 - \frac{\partial P_{Loss}}{\partial P_i}\right)} \quad (27)$$

where

$-\frac{\partial P_{Loss}}{\partial P_i}$ is the incremental transmission loss that can be calculated *by*

$$\frac{\partial P_{Loss}}{\partial P_i} = \frac{\partial}{\partial P_i}\left(\sum_{l=1}^{nl} P_l^2 \times R_l\right) \quad (28)$$

Equation (28) can be reformulated as equation (29) and equation (30).

$$\frac{\partial P_{Loss}}{\partial P_i} = \frac{\partial}{\partial P_i}\left(\sum_{l=1}^{nl}\left(\sum_{i=1}^{ni} A_{i,l} \times P_i\right)^2 \times R_l\right) \quad (29)$$

$$\frac{\partial P_{Loss}}{\partial P_i} = \left(2 \times \sum_{l=1}^{nl} A_{i,l} \times \left(\sum_{i=1}^{ni} A_{i,l} \times P_i\right) \times R_l\right) \quad (30)$$

where

P_i net injection at bus i (MW)

P_l power flow in line l (MW)

R_l line resistance (ohm).

The marginal loss pricing formulation requires Penalty factors and Delivery Factors. The Delivery factor of bus i can be calculated as in (31):

$$DF_i = \left(\frac{1}{PF_i}\right) = \left(1 - \frac{\partial P_{Loss}}{\partial P_i}\right) \quad (31)$$

3 Case Study

The case study presented in this paper show the results of the transmission costs distribution [Ferreira et al 2003], imputed to the loads or generators. In this study case all the produced power is commercialized by bilateral contracts for the period of one year. It is considered the pair generation/load, as are respectively the production and its associated load to one determined wheeling agent, or traditional utility.

The transmission network used in the case study is represented in Figure 11.2.

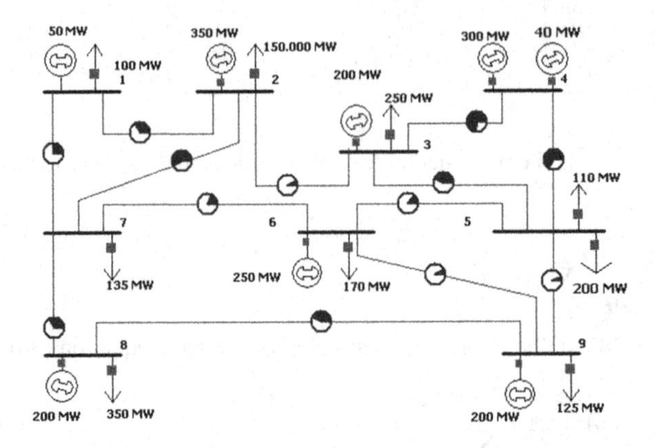

Fig. 11.2 Nine bus network

For this case study the transmission taxes imputed to the 20 bilateral transactions that are presented in the scenario presented in table 11.1 are calculated. These transactions are relative to the bilateral contracts carried through between generators and loads that integrate the network presented in figure 11.1.

Table 11.1 Bilateral Transactions

		L1	L2	L3	L5_1	L5_2	L6	L7	L8	L9
Ger	(MW)	100	150	250	110	200	170	135	350	125
G1	50	0	T2-50	0	0	0	0	0	0	0
G2	350	T1-100	0	0	0	0	T9-30	T12-20	T15-100	T18-100
G3	200	0	T3-100	0	0	0	0	T13-100	0	0
G4_1	300	0	0	0	T5-100	T7-160	T10- 40	0	0	0
G4_2	40	0	0	0	T6-10	0	0	T14-15	0	T19-15
G6	250	0	0	T4-250	0	0	0	0	0	0
G8	200	0	0	0	0	T8-40	T11- 100	0	T16-50	T20-10
G9	200	0	0	0	0	0	0	0	T17-200	0

To obtain the distribution of transmission costs were performed the following steps:

- The power flow in the transmission lines was evaluated in the simulator "Power World" Simulator. First, was been determinate the power flow at the base state of the system, presented on F_k, with all the loads and generators connected in the network and with no resistance lines consideration;

- To calculate the contribution due to each transaction *u* (pair generator/load), is removed the pair associated to it and in this situation the power flow is carried out with a new simulation.
- Analyzing the results of these two simulations allow to calculate the power flows in the $F_k(u)$ lines, associates to the transaction *u*. This process is repeated for all the transactions.
- Finally using the embedded methods, and knowing the contributions of each transactions in each line, is calculated $R(u)$ the tax to assign to each transaction *u*. The Matlab was the programming tool used to the implementation of the exposed methods.

3.1 Results

Table 11.2 Taxes imputed to each transaction ($)

T	PS	%	MW	%	B	%	UM	%	ZCF	%	DF	%
1	4133	6,3	6405	9,7	-4966	-7,6	3233	4,9	3566	5,4	3474	5,3
2	2066	3,1	-3186	-4,8	2597	4,0	1623	2,5	758	1,2	1256	1,9
3	4133	6,3	5926	9,0	51620	78,6	3846	5,9	4039	6,1	3664	5,6
4	10331	15,7	10329	15,7	-352	-0,5	15519	23,6	15185	23,1	13934	21,2
5	4133	6,3	10535	16,0	18618	28,3	3083	4,7	4235	6,4	3602	5,5
6	413	0,6	1061	1,6	1927	2,9	313	0,5	430	0,7	365	0,6
7	6612	10,1	16778	25,5	29270	44,5	4895	7,5	6723	10,2	5723	8,7
8	1653	2,5	-2963	-4,5	-6684	-10,2	1950	3,0	1285	2,0	1721	2,6
9	1240	1,9	-3087	-4,7	-16478	-25,1	1867	2,8	1076	1,6	1813	2,8
10	1653	2,5	-388	-0,6	-3053	-4,6	2416	3,7	1804	2,7	2502	3,8
11	4133	6,3	-18958	-28,9	-43603	-66,4	4609	7,0	589	0,9	3844	5,9
12	827	1,3	1561	2,4	-1954	-3,0	668	1,0	824	1,3	746	1,1
13	4133	6,3	13723	20,9	41679	63,4	5229	8,0	6546	10,0	5567	8,5
14	620	0,9	2573	3,9	5652	8,6	838	1,3	1076	1,6	939	1,4
15	4133	6,3	8668	13,2	-11341	-17,3	3971	6,0	4997	7,6	4440	6,8
16	2066	3,1	-9	0,0	5	0,0	3	0,0	1	0,0	2	0,0
17	8265	12,6	14433	22,0	24278	36,9	5163	7,9	5975	9,1	5514	8,4
18	4133	6,3	1412	2,1	-23819	-36,3	5385	8,2	5449	8,3	5479	8,3
19	620	0,9	1616	2,5	3541	5,4	836	1,3	1009	1,5	908	1,4
20	413	0,6	-724	-1,1	-1230	-1,9	259	0,4	140	0,2	213	0,3
CT	65707	100,0	65707	100,0	65707	100,0	65707	100,0	65707	100,0	65707	100,0

The results obtained for the taxes to be imputed to the 20 considered transactions are presented in table 11.2. This table presents, for each transaction, the taxes, in absolute and perceptual value. The methods considered are the following: Post-Stamp Method (PS), MW-Mile (MM), Base (B), Module or Use (MU), Zero Counterflow (ZCF) and Dominant Flow (DF).

The total cost of transmission is 65707 kEuro (CT). The calculation of this value is undertaken using the cost attributed C_k to each line, as follows:

$$CT = \sum_{all\ lines\ k} C_k = C_{1\text{-}2}+ C_{1\text{-}7}+ C_{2\text{-}3}+ C_{2\text{-}7}+ C_{3\text{-}4}+ C_{3\text{-}5}+ C_{4\text{-}5}+ C_{5\text{-}6}+ C_{5\text{-}9}+ C_{6\text{-}7}+ C_{6\text{-}9}+ C_{7\text{-}8}+ C_{8\text{-}9} = 65707 \text{ kEuro} \quad (32)$$

3.1.1 Comparison of the Taxes Imputed to the Transactions

With the taxes values presented in Table 11.2, it is possible to compare the imputed costs to some of the players that participate in the studied scenario. These costs are compared to the transactions, generators and loads that present similar values.

The transactions T1 and T15 correspond to the same transaction power (100MW); however they are taxed with different values by most tariff methods, as it can be seen in the graph of figure11.3.

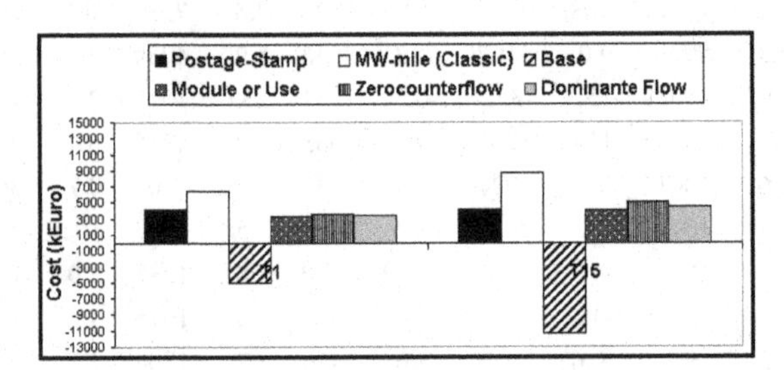

Fig. 11.3 Taxes Imputed to Transactions T1 and T15

With the exception of the Postage Stamp method, these two transactions pay different values because the distances between the involved loads and generators are different for the two transactions. The value taxed to T15 is higher than the value taxed to T1 because T15 uses the network in a more intensive way.

Considering transactions T2 and T16, each one corresponding to 50 MW, the values to impute are graphically presented in figure 11.4.

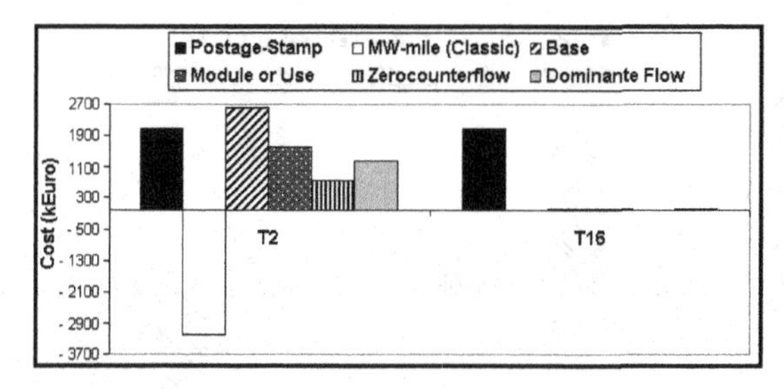

Fig. 11.4 Taxes Imputed to Transactions T2 and T16

Transaction T16 is taxed in almost all the methods with practically null values, compared with the total cost of transmission. The exception is the Post Stamp method where the transaction is taxed by 3, 14 % (2066 kEuro) of the total transmission cost. This happens because the load and the generator involved in transaction T16 (load L8 and generator G8) are located in the same bus.

By the analysis of the graph represented in figure 11.4, it can be verified that T2 is taxed with positive value in the methods Post-Stamp, Base, Module or Use, Zero Counterflow and Dominant Flow and with negative value by the method MW-mile Classic.

Although transactions T2 and T16 have the same contractual value of power, the difference of the value of the taxes imputed to these transactions is due to the fact that for transaction T2 the involved load and generator (load L2 and the generator G1) are not connected to the same bus as happens for transaction T16.

3.1.2 Comparison of the Taxes Imputed to the Generators

It can be easily concluded that with generators that inject the same power in the network can be taxed with different values. These values depends on the loads that the producers have contracts with, and on the effect of these contracts on the active power flows.

In order to analyze the results from the point of view of the generators, let us consider for example generators producing 200 MW. The taxes to impute to these generators (G3 and G8) are represented in the graphic of figure 11.5.

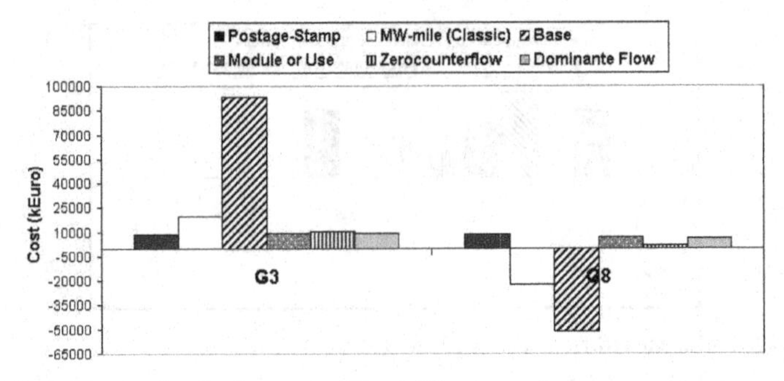

Fig. 11.5 Taxes Imputed to the Generators G3 and G8

Analyzing these results, it can be concluded that the methods for which the taxed values are closer are the Module or Use, Zero Counterflow and Dominant Flow. Although these methods do not give credit to the generators but they tax them with lower values. For example, the taxes calculated for generator G8 by these three methods do not give any credit to the generator but the taxes are lower than Post Stamp method.

Analyzing the graphic of table11.3, one can conclude that G3 always pays the same or more than G8; G8 receives a credit when MW-Mile and Base methods are used.

Table 11.3 Taxes Imputed to the Generators G3 and G8

	G3	G8
Postage-Stamp	8265	8265
MW-Mile	19649	-22654
Base	93299	-51512
Module or Use	9075	6822
Zerocounterflow	10584	2015
Dominante Flow	9231	5781

Generator G8 is involved in more transactions than generator G3, for the same total power transacted. Therefore, we can conclude, that, from the point of view of transmission taxes, it may be more advantageous to the generator to make several transactions instead of a single transaction. This allows that the caused power flows can be diversified avoiding possible high power flows that can be more heavily taxed.

3.1.3 Comparison of the Taxes Imputed to the Loads

The aim of comparing the taxes imputed to the loads is to verify which approach can present more benefits if the utilization of the transmission network is paid by the consumers.

The values presented in table 11.2 were obtained adding the taxes imputed to all the transactions associated each load. For example, the tax value calculated for load L5_1 has been obtained adding the tax values of the transactions T5 and T6 in which load L5_1 is involved. The values of the taxes concerning loads L1 and L5_1are presented in figure 11.6. Loads L1 and L5_1 have slightly different values of active power (load L5_1 has a value of active power 10% greater that load L1). These loads are involved in the following transactions:

- Load L1 contracts 100 MW to generator G2;
- Load L5_1 contracts 100 MW to generator G4_1 and 10 MW to generator G4_2.

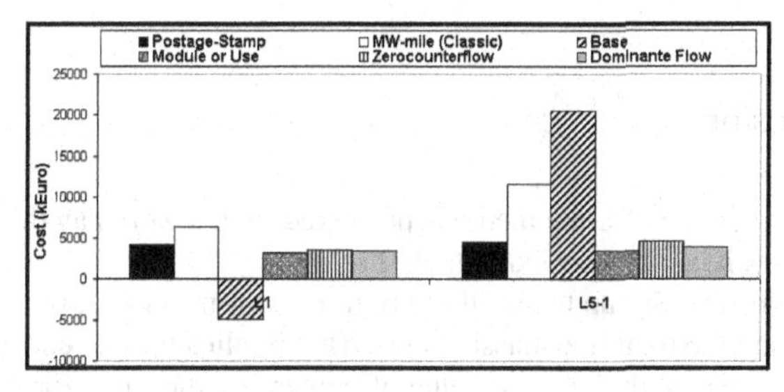

Fig. 11.6 Taxes Imputed to Loads L1 and L5_1

Analyzing the values presented in the graph of figure 11.6, it can be concluded that for all the six studied methods load L5_1 always pays more than load L1. In table11.4 the values of the taxes that are in the column identified with (L1+10%L1) correspond to auxiliary calculations that determine the value of the taxes imputed to L1 increased by 10%.

Table 11.4 Taxes Imputed to the Loads L1 and L5_1

	L1	L1+10%L1	L5_1
Postage-Stamp	4133	4546	4546
MW-Mile	6405	7046	11596
Base	-4966	-5463	20545
Module or Use	3233	3556	3396
Zerocounterflow	3566	3922	4665
Dominante Flow	3474	3821	3967

The taxes imputed to load L1 vary between a situation in which the load receives 4966 kEuro (7,56% of the total), calculated by the method Base and a situation in which it pays a maximum value of 6405 kEuro (9,75% of the total cost) taxed by the Classic MW-mille method. In the case of load L5_1 the extreme payment values vary from a minimum of 3396 kEuro (5,17%), calculated by the method Module or Use, and a maximum of 20545 kEuro (31,27%) , calculated by the method Base.

It can also be verified that the taxes calculated by the method Postage-Stamp increase proportionally with the increase of the value of the power contracted by the load. For the remaining methods the values taxed to L5_1 increase more than 10% of the value taxed to L1, as can be analyzed in table 11.4.

Acknowledgments The authors would like to acknowledge FCT, FEDER, POCTI, POSI, POCI, POSC for their support to R&D Projects and GECAD Unit. We also thank GECAD researchers, especially Bruno Canizes, João Soares and Tiago Sousa for their help in the simulations undertaken for the illustrative case studies.

4 Conclusion

All the taxes calculated for the methods presented in this work pay the total cost of transmission, as it can be verified in table 11.2.

The method Postage-Stamp taxes all the transactions in function of the contracted power value and the total transmission cost. This implies for one electrical system, that all transactions with the same value of power pay the same for the use of the transmission network, even when d not use it. For example, a transaction between producers and consumers on the same bus, that do not need use the electrical lines for physical concretization of the transaction, is taxed in the same way that another one, that carries through between producers and consumers who are in different bus, needing therefore to use the network.

In relation to the method Base we conclude that this method allow that the agents receive when the taxes are negatives. When they are taxed positively agents must pay very high values. We can conclude that this method is not advantageous to the transactions, because it provokes great difference in the values taxed to the transactions with equal power transacted.

The methods MW-Mile, Module or Use, Zero Counterflow and Dominant Flow always tax the agents with the positive value. Although these values are smaller comparing identical transactions and they are more uniform. These methods tax in function of the transmission impacts in the active power flows.

References

Abhyankar R, Soman A and Khapard A (2006) Optimization Approach to Real Power Tracing: An Application to Transmission Fixed Cost Allocation, IEEE transactions on power systems, vol. 21, no. 3.

Bialek J (1997), Topological generation and load distribution factors for supplement charge allocation in transmission open access, IEEE Trans Power Syst. 12 (1).

Conejo J, Contreras J, Lima A and Padilha-Feltrin A (2007) Zbus transmission network cost allocation, IEEE Trans. Power Syst. 22 (1), 342–349.

Dicorato, L'Abbate A, Minoia A and Trovato M (2003) Electric Power Systems Research, Volume 66, Issue 2, Pages 179-186 M.

Ferreira J, Vale Z and Puga R (2009) Nodal Price Simulation in Competitive Electricity Markets, Proceedings of 6th International Conference on the European Electricity Market, EEM 09, Leuven, Belgium.

Ferreira j, Vale Z, Vale A and Puga R (2003) Cost of Transmission Transaction: Comparison and Discussion of Used Methods International Conference on Renewable Energies and Power Quality (ICREPQ'03) Vigo

Galetovic A and Palma-Behnke R (2008) Can generalize distribution factors lead to 'objective' transmission toll allocations? Some lessons from the recent Chilean experience, Energy Economics, Volume 30, Issue 2, Pages 249-270.

Kirschen S, Allan N and Strbac G (1997) Contribution of individual generators to loads and flows, IEEE Trans. Power Syst. 12 (1).

Madiji M, Ghazizadeh S and Afsharnia S (2008) A novel approach to allocate transmission embedded cost based on MW-Mile method under deregulated environment, IEEE Electrical Power & Energy Conference.

Rothwell G and G6mez T (2003) Electricity Economics _Regulation and Deregulation, Book by Institute of Electrical and Electronics Engineers.

Sood R, Padhy P and Gupta O (2007) Deregulated model and locational marginal pricing, Electric Power Systems Research, Volume 77, Issues 5-6, Pages 574-582.

Tan S, Yun L and Lo L (2008) A more transparent way of financial settlement for congestion cost in electricity markets, Third International Conference on Electric Utility Deregulation and Restructuring and Power Technologies.

Zhaoxia J and Fushuan W (2005) Discussion on the Proving of Proportional Sharing Principle in Electricity Tracing Method, IEEE/PES Transmission and Distribution Conference & Exhibition: Asia and Pacific Dalian, China.

Computational Intelligence Applications for Future Power Systems

Zita Vale[1], Ganesh K. Venayagamoorthy[2], Judite Ferreira[1] and Hugo Morais[1]

[1]GECAD – Knowledge Engineering and Decision-Support Research Group of the Institute of Engineering, Polytechnic of Porto (ISEP/IPP), Rua Dr. António Bernardino de Almeida, 4200-072 Porto, Portugal *(email: {zav, mju, hgvm}@isep.ipp.pt)*

[2]Real-Time Power and Intelligent Systems Laboratory, Missouri University of Science and Technology, Rolla, MO 65409-0249, USA *(email: gkumar@ieee.org)*

Abstract Power system planning, control and operation require an adequate use of existing resources as to increase system efficiency. The use of optimal solutions in power systems allows huge savings stressing the need of adequate optimization and control methods. These must be able to solve the envisaged optimization problems in time scales compatible with operational requirements. Power systems are complex, uncertain and changing environments that make the use of traditional optimization methodologies impracticable in most real situations. Computational intelligence methods present good characteristics to address this kind of problems and have already proved to be efficient for very diverse power system optimization problems. Evolutionary computation, fuzzy systems, swarm intelligence, artificial immune systems, neural networks, and hybrid approaches are presently seen as the most adequate methodologies to address several planning, control and operation problems in power systems. Future power systems, with intensive use of distributed generation and electricity market liberalization increase power systems complexity and bring huge challenges to the forefront of the power industry. Decentralized intelligence and decision making requires more effective optimization and control techniques techniques so that the involved players can make the most adequate use of existing resources in the new context.
The application of computational intelligence methods to deal with several problems of future power systems is presented in this chapter. Four different applications are presented to illustrate the promises of computational intelligence, and illustrate their potentials.

1 Power Systems – Present and Future

Traditionally the power sector was based on vertically integrated utilities (generation, transmission, distribution, and retail) which were, in many countries, com-

A. Madureira et al. (eds.), *Computational Intelligence for Engineering Systems: Emergent Applications*, Intelligent Systems, Control and Automation: Science and Engineering 46, DOI 10.1007/978-94-007-0093-2_12, © Springer Science + Business Media B.V. 2011

pletely or mostly state owned. In recent years remarkable changes are taking place, with the power industry moving toward a new structure of vertical disintegration. The new organization is mainly based on liberalized electricity markets that promote competition. Generators compete to sell energy in the pool and/or trough bilateral contracts and retailers compete to attract more electricity consumers [Shahidehpour et al 2002, Vale 2009].

Under this new structure, with competitive trading between wholesale market participants, operating functions are clearly defined and divided into market operation functions and system operation functions. Market operation functions are associated with energy trading, scheduling, and settlement of energy transactions in different time horizons and are assured by the market operator. System operation functions are directly related with the physical operation and restrictions of the power system, dealing with operation and control of the bulk power system and the need to assure load and security requirements. This must consider the requirement of a real time dispatch to balance supply and demand, the need to maintain system security and requires ancillary services and transmission congestion management [Shahidehpour et al 2002, Miller and Malinowski 1994, Vale et al 2010].

With electricity market liberalization on the way, the power industry is also experiencing other huge changes, mainly related with the need to limit this high polluting industry impact on the environment. This caused energy policies around the world, with emphasis on the European Union [European Communities 2008] and the United States [US-DoE 2009a], to favor energy efficiency and renewable energy sources use increase. Due to the distributed nature of such energy sources, such as wind and sun, this has led to increasing levels of distributed generation of electrical energy. Besides reducing greenhouse gas emissions and environmental impact of electricity generation, this decentralization trend, when compared with the traditional centralized generation, can have some relevant advantages such as cost reduction, higher service quality levels (mainly in incident situations), and losses reduction. However, taking advantage of distributed generation requires changes in electrical networks planning and operation and distributed generation increase puts complex problems to present networks operation [Ilic et al 2007].

The new reality and the impossibility of accommodating intensive levels of distributed generation with the presently used paradigms led to a new paradigm known as smart grids. The smart grid can be viewed as a digital upgrade of the existing electricity infrastructure to allow for dynamic optimization of current operations as well as incorporate dynamic gateways for alternative sources of energy production [Venayagamoorthy 2009].

Smart grids, sometimes referred to as Intelligent Grid/Intelligrid and FutureGrid [Venayagamoorthy 2009], must have certain basic functions, as indicated in the Energy Independence and Security Act of 2007, including [Venayagamoorthy 2009, Energy Independence and Security Act 2007,US-DoE 2008, US-DoE 2009b]:

Self-healing – capability to recover from faults and restore functionality;
Fault tolerance – resist attacks;
Allow for integration of all forms of energy generation and storage options including plug-in vehicles;
Allow for dynamic optimization of grid operation and resources with full cyber-security;
Allow for incorporation of demand-response, demand-side resources and energy efficient resources;
Allow electricity clients to actively participate in the grid operations by providing timely information and control options;
Improve reliability, power quality, security and efficiency of the electricity infrastructure.

Although the discussion around smart grids has turned mainly around smart metering [Tai and HÓgáin 2009] for a long time, presently smart grid discussions refer to a much wider set of requirements, involving issues from conceptual design to equipment and software applications [Giri et al 2009, Rahman 2009].

One can say that a smart grid is an infrastructure able to accommodate all centralized and distributed energy resources (DER) [Morais et al 2010], including intensive use of renewable and distributed generation, storage, plug-in vehicles (electric vehicles –EV, hybrid electric vehicles - HEV) [Saber and Venayagamoorthy 2010], and demand response (DR), seeing consumers as active players [Vale et al 2009a], in the context of a competitive business environment [Azevedo et al 2007, Praça et al 2003]. Smart grids should provide the required means to efficiently manage all these energy resources (ER) and to decentralize control and intelligence throughout the power system.

This vision requires that smart grids provide the means so that a wide range of diverse scale players can fairly act in the energy business. These include producers and consumers from large to small and even micro scale and diverse types of aggregators.

In order to carry out the smart grid functions mentioned above, advanced monitoring, forecasting, decision making, control and optimization algorithms are required. These algorithms must be fast, scalable and dynamic to be able to deal with future power systems requirements. The importance of computational intelligence based algorithms in the context of present and future power systems is discussed in this chapter.

2 Computational Intelligence Methods in Power Systems

Computational intelligence (CI) is the study of adaptive mechanisms to enable or facilitate intelligent behavior in complex, uncertain and changing environments. These adaptive mechanisms include bio-inspired and artificial intelligence paradigms that exhibit an ability to learn or adapt to new situations, to generalize, abstract, discover and associate [Venayagamoorthy 2009, Engelbrecht 2007]. CI most used paradigms include neural networks, evolutionary computation, fuzzy systems, swarm intelligence and artificial immune systems. These paradigms can be combined to form hybrids as shown in Fig.12.1 resulting in Neuro-Fuzzy systems, Neuro-Swarm systems, Fuzzy-PSO systems, Fuzzy-GA systems, Neuro-Genetic systems, etc [Venayagamoorthy 2009].

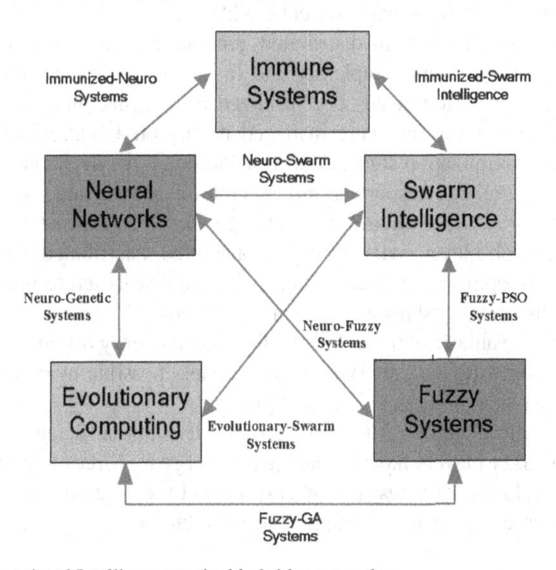

Fig.12.1 Computational Intelligence typical hybrids approaches

The main features of the CI approach that is attractive and promising for future power systems includes their ability to:

- Carry out behavioral modeling;
- Carry out fast and accurate decision making;
- Identify/predict system nonlinear dynamics;
- Provide robust/optimal/adaptive coordinated control on nonlinear and non-stationary systems;
- Perform complex and large-scale optimization;
- Provide immunized solutions;
- Self-heal;
- Scale-up;

Optimization studies are extensively used in power system planning (e.g. for distribution network planning) and operation (e.g. for economic dispatch). In recent years, computational intelligence methods have been used for optimization purposes in many different applications in power systems [Lee and El-Sharkawi

2008]. The adequacy of these methods is due to the characteristics of the target problems, this adequacy being even more evident in future power systems context.

Today's interconnected power systems are complex adaptive systems under semi-autonomous control. A power system is spatially and temporally complex, nonconvex, nonlinear and non-stationary with uncertainties at many levels. Its efficient, reliable and safe operation and control is a challenge today as seen from several blackout situations such as the 2003 Northeast American Blackout [Venayagamoorthy 2009, Amin and Wollenberg 2005].

The complexity grows with distributed generation increase. To overcome the challenges brought by this complexity and to enable the participation of market players, the power system needs decentralized intelligence and decision making. Traditionally, power systems were managed mostly on a centralized way, giving place to global optimization studies that considered all the available resources (including network related resources and generation plants) and load forecasting. Operation in a liberalized environment changes the way optimization studies are undertaken as each player goals are to be considered. Distributed generation must be considered in optimization studies but the use of distributed resources has to respect contractual relationships among market players.

Taking full advantage of the available distributed energy resources in a competitive environment with a diversity of players is only possible by successfully leading to practice two concepts [Vale et al 2009b]:

a) Real decentralization of decision-making. It is not sufficient that the power system and market players have the adequate energy resources to make them able of providing valuable services. It is necessary that they are able to undergo decision making processes that give support to strategic behavior to accomplish their goals;

b) All available resources should be considered in planning and operation studies. What is important is not who owns the relevant resources to solve a problem. What is really important is if the player who is undertaking the study has permission to use those resources. This permission has to take into account resource ownership and all the commercial relationships that can allow one entity to use others' resources (provided the corresponding value is fairly paid to the resource owner, on a market based approach).

This approach changes the way optimization studies should be undertaken. In the new context, optimization is used by each player taking into account that each resource owner must be fairly remunerated when supplying a service and that market mechanisms can allow (at a price) one player to use others resources. Resource management takes a much more intelligent perspective and available resources can be fully used for solving power system planning and operation issues. Intelligent information processing is needed everywhere a critical decision is made for performance control, locally or globally, requiring a large number of distributed sensors and actuators. The traditional methods for power system modeling, control and optimization need to be rethought in order to incorporate intelligent techniques capable of rapid adaptation, having dynamic foresight, being fault-tolerant and robust to disturbances and randomness [Venayagamoorthy 2009].

This chapter presents some illustrative examples on how computational intelligence can adequately address the new challenges of future power systems optimization, control and operation.

3 Computational Intelligence applications in Power Systems – Some Examples

3.1 Ancillary Services Dispatch using a Genetic Algorithm Approach

In electricity markets, ancillary services (AS) can be seen as a set of products separated from the energy transactions. Adequate procedures and methodologies to determine AS needs as well as the ways they are obtained and priced are required in a competitive environment [Zita et al 2010, Pereira et al 2004, Vale et al 2009c, Papalexopoulos 2002]. Modern meta-heuristics approaches can be applied to solve the dispatch of AS.

In this section, the application of a genetic algorithm (GA) based approach for AS dispatch is presented, using an illustrative example. The formulation presented in this paper considers Regulation Down, Regulation Up, Spin Reserves and Non-spin Reserves services.

The Ancillary Services Dispatch Optimization can be formulated by (1) to (3) for Regulation Down and by (4) to (9) for Regulation Up, Spin Reserves and Non-spin Reserves. Regulation down dispatch is done independently from the other services as it does affect the maximum available capacity of market players [Zita et al 2010, Pereira et al 2004, Vale et al 2009c, Papalexopoulos 2002]:

$$MIN\sum_{i=1}^{N}\left(PR_{i,k}+p\cdot PE_{i,k}\right)\cdot X_{i,k} \qquad (k=1) \tag{1}$$

Subject to:

$$\sum_{i=1}^{N} X_{i,1}=Q_1 \tag{2}$$

$$0\leq X_{i,1}\leq C_{i,1} \ and \ X_{i,1} \ integer \tag{3}$$

$$MIN\sum_{k=2}^{4}\sum_{i=1}^{N}(PR_{i,k}+p.PE_{i,k})*X_{i,k} \tag{4}$$

Subject to:

$$\sum_{i=1}^{N}X_{i,2}=Q_2 \tag{5}$$

$$\sum_{i=1}^{N}X_{i,3}=Q_3 \tag{6}$$

$$\sum_{i=1}^{N}X_{i,4}=Q_4 \tag{7}$$

$$\sum_{k=2}^{4}X_{i,k}\leq Cm\acute{a}x_i \quad \text{I=1,...,N} \tag{8}$$

$$0\leq X_{i,k}\leq C_{i,k} \text{ and } X_{i,k} \text{ integer} \tag{9}$$

where N is the total number of bids, i is the bid index (i=1,2,….,N), k is the Ancillary service index (1 for regulation down, 2 for regulation up, 3 for spin reserve, and 4 for non-spin reserve), Q_k is the total capacity requirement for ancillary service k, $Cmax_i$ is the maximum capacity of bid i, C_{ik} is the capacity of bid i for ancillary service k, $X_{i,k}$ is the accepted capacity of resource i for ancillary service k, $PR_{i,k}$ is the price bid for reserve capacity of bid i for ancillary service k, $PE_{i,k}$ is the price bid for reserve energy of bid i for ancillary service k, and p is the estimated probability of using energy reserves acquired for ancillary services.

Table 12.1 presents the data for this case study. These data include the ancillary services bids for a set of 10 players and the values of the required power reserve for each service, which are displayed in the last row of Table 12.1. Quantities (Qt) are expressed in MW, the total capacity requirement (Rq) in MW, reserve prices (PR) in mu/MW and energy prices (PE) in mu/MWh, where mu stands for monetary units.

Table 12.1 Ancillary service bids and requirements

	Ancillary Services												
	Max.	Regulation Down			Regulation Up			Spin Reserve			Non-Spin Reserve		
Bids	Power (MW)	Qt (MW)	PR (mu/MW)	PE (mu/ MWh)	Qt (MW)	PR (mu/ MW)	PE (mu/ MWh)	Qt (MW)	PR (mu/ MW)	PE (mu/ MWh)	Qt (MW)	**PR (mu/ MW)**	**PE (mu/ MWh)**
1	100	70	10.0	50.0	80	15.0	25.8	10	5.0	45.6	95	7.0	30.6
2	80	80	8.0	40.0	70	10.0	45.0	55	9.0	50.6	45	4.0	40.0
3	90	55	8.0	38.8	80	8.2	50.8	88	8.5	61.8	70	6.0	45.8
4	79	60	4.0	59.6	50	5.0	60.6	30	7.2	70.8	41	4.6	35.8
5	100	100	3.5	76.4	65	7.0	50.3	24	4.0	48.0	45	9.0	40.0
6	90	20	9.0	47.4	50	14.6	33.4	80	8.0	39.4	18	11.0	49.4
7	110	40	7.0	50.0	98	8.5	30.4	50	7.3	48.0	24	10.5	38.0
8	90	10	4.8	60.0	100	10.6	50.0	50	6.6	57.6	80	4.0	34.6
9	75	40	9.0	35.0	110	6.5	35.0	10	4.3	54.4	80	5.3	35.4
10	85	15	10.0	40.0	40	4.0	40.0	49	9.0	35.4	90	5.0	40.0
Rq	750	250			200			150			150		

Table 12.2 presents the crossover, mutation, select operator and number of generations used in the GA algorithm.

Table 12.2 GA tuning parameters

Attribute	Description	Used value Regulation Down	Used value Regulation up, Spin reserve and non-spin reserve
ε	Epsilon indicates the threshold to find the best solution	10–10	10–10
Crossover operator	Arithmetic Crossover	25 crossover operations per generation	12 crossover operations per generation
Mutation operator	Boundary Mutation	10 mutations per generation	70 mutations per generation
Select operator	Normalized geometric distribution	Probability of 1.00	Probability of 1.00
Select operator 2	Roulette method	-	-
Initial population	Population size	100	250
Number of generations	Number of generations applied to start population	100	80

Table 12.3 presents a synthesis of the obtained results. These results consider 0, 50 and 100% probability for the effective use of the reserve service (p=0.0, p=0.5 and p=1.0). Total represents the capacity requirement, MCP is the market clearing price and Final Cost is the total reserve cost, considering the set of four AS, including reserve capacity and energy costs. For p=0.0, the reserve energy is not used so the Final Cost equals the Reserve Capacity Total Cost.

Table 12.3 Ancillary services dispatch results' by GA

GA	Regulation Down			Regulation Up			Spin Reserve			Non-Spin Reserve			Four Ancillary Services Set		
	p														
	0.0	*0.5*	*1.0*	*0.0*	*0.5*	*1.0*	*0.0*	*0.5*	*1.0*	*0.0*	*0.5*	*1.0*	*0.0*	*0.5*	*1.0*
Total (MW)	250	250	250	200	200	200	150	150	150	150	150	150	750	750	750
MCP (mu/MW)	8	10	10	8.5	8.5	10	8.5	9	9	4.6	7	7			
Reserve Capacity Total Cost (mu)	2000	2500	2500	1700	1700	2000	1275	1350	1350	690	1050	1050	5665	6600	6900
Final Cost (mu) (Energy cost included)	p=0.0	5665.00													
	p=0.5	20555.75													
	p=1.0	35483.00													

The use of a GA approach allows achieving solutions that, although being slightly more costly than the optimal solution, are determined in lower computing times. Table 12.4 presents the computing times required to solve AS dispatch for three case studies with 10, 15 and 22 bids. It is important to note that the computing complexity depends more on the number of the unique bid combinations than on the number of bids. These results and other results obtained in more extensive simulations that we have undertaken clearly show that the GA approach is less time consuming, with special relevance for larger problems. This means that a GA approach can be used to overcome the time consuming problems of traditional optimization approaches whenever large, combinatorial problems must be solved within short time limits.

Table 12.4 LP and GA times comparison with 10, 15 and 22 bids

Time (s)						
Bids	LP			GA		
	p=0	p=0.5	p=1	p=0	p=0.5	p=1
10	6.10	6.72	7.05	2.49	2.52	2.55
15	4.84	5.23	5.78	2.50	2.60	2.57
22	6.13	6.97	7.60	3.30	3.43	3.51

3.2 Reactive Power Management using a PSO Approach

The main objective of the case study presented in this sub-section is to adjust the reactive power of an already known generation scheduling plan, using the available transformer taps and capacitor banks. The purpose is to minimize active power loss and to assure that the bus voltages lie within the a priori defined limits, subjected to the power system technical constraints. The considered constraints are the ones imposed by the active and reactive power flow equations and by power generation constraints. All constraints that have been represented in the objective function are weighted with penalty factors. The IEEE 14-bus system has been used as test case to illustrate the application of the proposed algorithm.

The used system is shown in Fig.12.2 and the network parameters have been extracted from [Zimmerman and Gan 1997, 29, Kodsi and Canizares 2003]. The network consists of 5 generator-buses, 11 load-buses and 20 branches.

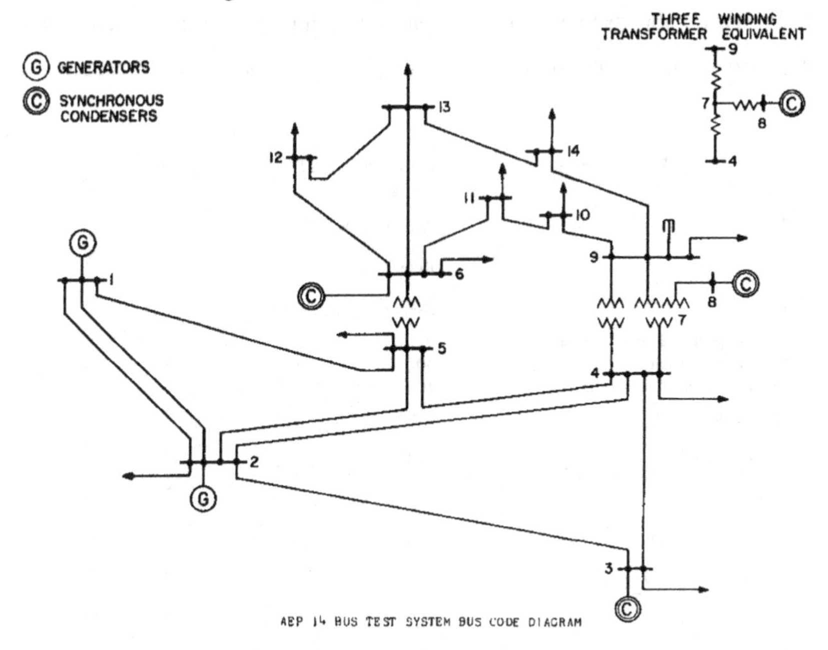

Fig. 12.2 IEEE 14-bus power system [Kodsi and Canizares 2003]

Particle swarm optimization can be used to address this problem, having been demonstrated to be effective in addressing several power system problems [Lee and El-Sharkawi 2008, Grant et al 2008, Chaturvedi 2008, Wu et al 2008]. The movement rule of PSO is defined by three strategic parameters: inertia, memory and cooperation and does not consider selection or mutation operators [Lee and El-Sharkawi 2008, Mendes 2004]. Populations' topologies are determinant for the performance of particle swarm. An evolutionary variant of PSO can be used, leading to an evolutionary approach in which mutation of the strategic parameters and selection by stochastic tournament of particles passing to the next generation are applied [Lee and El-Sharkawi 2008, Mendes 2004, Miranda et al 2008]. This new approach can be referred to as evolutionary particle swarm optimization (EPSO) [Lee and El-Sharkawi 2008, Miranda et al 2008], several distribution functions can be used in the evolutionary process [Hang and Wang 2007, Lee 2007].

In each iteration of EPSO the parameters are changed using a mutation function and accordingly to a mutation factor (that can vary between 0 and 1). In the present case study the parameters have been changed using a uniform random mutation function and the mutation factor used was 0.75 as can be seen in Table 12.5. A high value of the mutation factor can result in divergent solutions and a low value can result in the stagnation of solution evolution. Table 12.5 depicts the parameters used for this case study in the PSO and EPSO approaches.

V_i is a variable that represents the voltage magnitude at bus i in p.u. The variable tap_i is the tap ratio of transformer t. Q_{Ci} is the reactive power capacitor at bus i.

Table 12.5 Parameters used in the PSO and ESPO approaches

Parameter	Value	PSO	EPSO
V_i	Refer to Table 6	X	X
tap_t	Refer to Table 6	X	X
Q_{Ci}	Refer to Table 6	X	X
Search Space (number of variables)	9	X	X
Number of Particles	40	X	X
Number of movement iterations	100	X	X
Max velocity of V_i	0.1	X	X
Min velocity V_i	-0.1	X	X
Max velocity of tap_t	0.1	X	X
Min velocity tap_t	-0.1	X	X
Max velocity of Q_{Ci}	18	X	X
Min velocity Q_{Ci}	-18	X	X
Mutation function	Uniform Rand Mutation		X
τ mutation factor	0.75		X

Table 12.6 shows the type and the limits of the decision control variables used in this case study. It can be seen that tap and Qc are discrete variables with defined ranges while V is a continuous variable.

Table 12.6 Decision variables description

Control Variables	Minimum Value (p.u.)	Maximum Value (p.u.)	Variable State
V_2	0.9	1.1	Continuous
V_3	0.9	1.1	Continuous
V_6	0.9	1.1	Continuous
V_8	0.9	1.1	Continuous
$tap_{4\text{-}7}$	0.9	1.1	Discrete – Step 0.01
$tap_{4\text{-}9}$	0.9	1.1	Discrete – Step 0.01
$tap_{5\text{-}6}$	0.9	1.1	Discrete – Step 0.01
Qc_9	0	0.18	Discrete – Step 1
Qc_{14}	0	0.18	Discrete – Step 1

Table 12.7 display the EPSO and PSO data results: voltage magnitude, angle, active and reactive generated power per bus and the reactive power to be compensated with capacitors.

Table 12.7 PSO data results

Bus	V		Voltage angle		Generated Power				Reactive Power Compensation	
N°	p.u.		°		P_G		Q_G		MVar	
	EPSO	PSO	EPSO	PSO	EPSO	PSO	EPSO	PSO	EPSO	PSO
1	1.0600	1.0600	0	0	232.2873	232.3601	-26.3816	-26.6404	-	-
2	1.0469	1.0477	-4.9844	-5.0062	40	40	30.3271	34.0536	-	-
3	1.0185	1.0215	-12.7107	-12.7688	0	0	26.2432	30.0196	-	-
4	1.0296	1.0281	-10.4031	-10.4003	0	0	0	0	-	-
5	1.0338	1.0312	-8.9103	-8.8647	0	0	0	0	-	-
6	1.0557	1.0377	-14.6054	-14.5272	0	0	28.6677	14.7913	-	-
7	1.0515	1.0491	-13.5989	-13.6878	0	0	0	0	-	-
8	1.0817	1.08	-13.5989	-13.6878	0	0	18.564	18.9666	-	-
9	1.0454	1.0421	15.2539	-15.3931	0	0	0	0	18	18
10	1.0397	1.0338	-15.4333	-15.5476	0	0	0	0	-	-
11	1.0441	1.0323	-15.1559	-15.1888	0	0	0	0	-	-
12	1.0422	1.0264	-15.4934	-15.4686	0	0	0	0	-	-
13	1.0388	1.025	-15.6209	-15.676	0	0	0	0	-	-
14	1.0339	1.0354	-16.6748	-17.0386	0	0	0	0	6	12

Table 12.8 displays the tap ratio variables obtained using PSO and EPSO.

Table 12.8 Taps variables

tap	PSO	EPSO
$tap_{4\text{-}7}$	0.96	0.90
$tap_{4\text{-}9}$	1.04	0.90
$tap_{5\text{-}6}$	0.90	1.04

Table 12.9 displays the PSO and EPSO run time figures as well as the best fitness found after 100 runs. It can be concluded that introducing mutation in strategic parameters of the PSO classical method it is possible to achieve a best solution with little CPU time addition.

Table 12.9 Performance of PSO and EPSO

Technique	Mean run time	Best fitness value found in 100 runs
PSO	22	13.3601
EPSO	25	13.2873

3.3 Vehicle-to-Grid (V2G) and Grid-to-Vehicle (G2V) Scheduling using PSO

Future power systems will have capabilities where plug-in vehicles can charge from them or discharge to them. An intelligent method for scheduling optimal usage of energy stored in these vehicles is necessary. The batteries on these vehicles can either provide power to the grid when parked (usually 95% of the time, known as the vehicle- to-grid (V2G) concept or take power from the grid to charge the batteries on the vehicles, known as the grid-to-vehicle (G2V). Determining optimal time schedules to charge or discharge in order to maximize profits to vehicle owners while satisfying system and vehicle owners' constraints has been studied by one of the authors for various power system problems including participation unit commitment, economic dispatch, and reactive power. In order to optimal schedule the vehicles binary particle swarm optimization was applied in [Hutson et al 2008]. Price curves from the California ISO database are used in this study to have realistic price fluctuations. Different fleets of vehicles are used to approximate varying customer base and demonstrate the scalability of parking lots for V2G. The results for consistency and scalability are included. Fig.12.3 shows a typical parking lot for plug-in electric vehicles connected to the utility grid. Table 12.10 shows the profit made by the vehicle owners based on a PSO scheduling on a given day. More details on this work are given in [Hutson et al 2008].

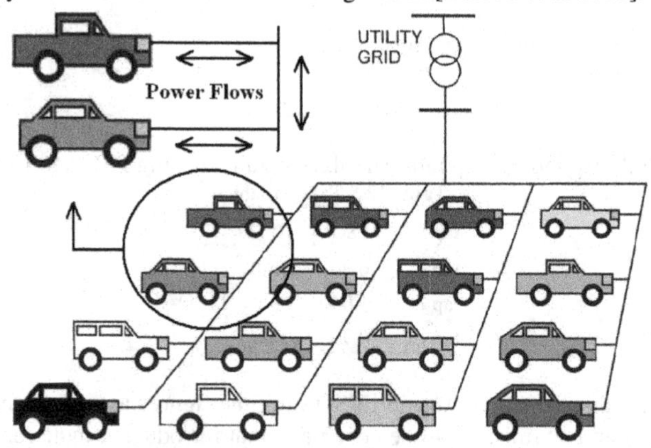

Fig. 12.3 A typical parking lot with plug-in electric vehicles [Venayagamoorthy 2009]

Table 12.10 Vehicles owners profits

# of Vehicles	Case Study	Power into Lot (MWh)	Power out of Lot (MWh)	Net Power Out (MW)	Profit
50	CS1	0.0089	0.1131	0.1042	$11.41
	CS2	0.3492	0.3421	-0.0072	$19.09
500	CS1	0.0984	1.2533	1.1549	$128.42
	CS2	3.5167	3.8271	0.3104	$234.22

5000	CS1	1.0359	12.1769	11.1401	$1223.49
	CS2	31.9632	35.2408	3.2777	$2200.40

V2G can reduce dependencies on small expensive units in the existing and future power systems as energy storage that can decrease running costs. It can contribute to efficient management of load fluctuation, peak load; however, it increases spinning reserves and reliability. As number of gridable vehicles in V2G is much higher than small units of existing systems, unit commitment (UC) with V2G is more complex than basic UC for only thermal units. Particle swarm optimization was proposed in [Saber and Venayagamoorthy 2010] to solve the V2G scheduling problem, as PSO can reliably and accurately solve complex constrained optimization problems easily and quickly without any dimension limitation and physical computer memory limit.

3.4 Wide Area Monitoring and Control Systems (WAMCS)

Neural networks are known to be universal approximators. Many papers published by one of the authors of this chapter have demonstrated their capabilities for to identify the changing dynamics of synchronous machines and multimachine power systems from moment to moment, and based on the identified dynamics, appropriate control signals can be generated using a second neural network to minimize overshoots, steady state errors and ensure fast settling times during disturbances and changes in operating conditions.

Simultaneous recurrent neural networks (SRN) are powerful tools for prediction based on internal recurrence and have been shown to overcome the varying time delays in the communication channels for wide area monitoring and control applications. The real-time implementation of a two-area four machine power system shown in Fig.12.4 was carried out at one of the author's laboratory – http://rtpis.org. Despite varying time delays, a system damping higher than that provided by local controllers (see Fig.12.5) can be guaranteed with wide area controllers (WAC) over a significant range of delays [Ray and Venayagamoorthy 2008]. This is possible based on the successful multi-step prediction capabilities of SRNs.

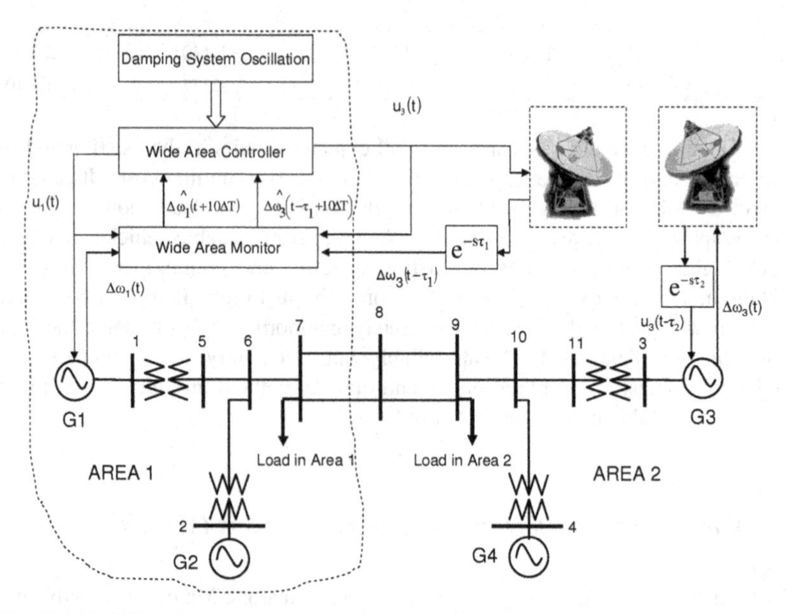

Fig. 12.4 WAMCS with communication delays considered in the design [Ray and Venayagamoorthy 2008]

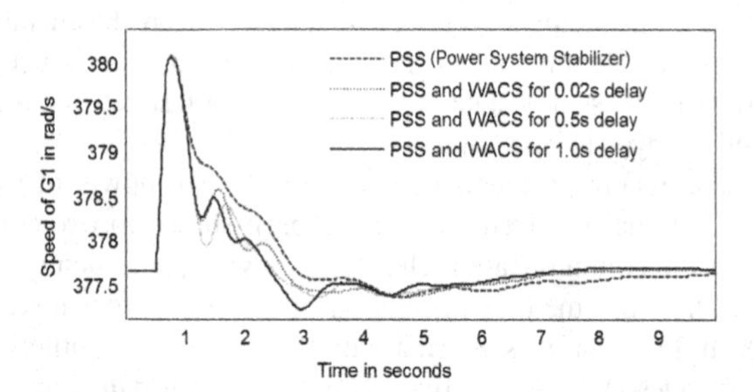

Fig. 12.5 Speed of generator G1 during a disturbance with different communication delays

Wide-area coordinating control is becoming an important issue and a challenging problem especially when renewable sources of energy such as wind farms and solar farms are integrated into the power grid. An optimal wide-area monitor based on a radial basis function (RBF) neural network has been developed in [Qiao et al 2008] to identify the input-output dynamics of the nonlinear power system. Its parameters are optimized through particle swarm optimization. Based on the wide area monitor (WAM), a WAC is then designed by using the dual heuristic programming (DHP) method and RBFs, while considering the effect of signal transmission delays. The WAC operates at a global level to coordinate the actions of local power system controllers including those on the wind farm. Each local controller communicates with the WAC, receives remote control signals from the

WAC to enhance its dynamic performance, and therefore helps improve system-wide dynamic and transient performance.

4 Conclusions

The complexity of future power systems is much more than that of the traditional power grid as time-varying sources of energy and new dynamic loads are integrated into it. Advanced intelligent techniques are required to handle the smart grid operation in an efficient and economical manner. Renewable sources of energy which are known today as not dispatchable will need accurate forecasting tools such as CI to become dispatchable tools. The potentials of computational intelligence paradigms have been demonstrated for various challenges in the traditional power system. Such techniques are promising solutions to deliver the expectations of future power systems.

Acknowledgments The authors would like to acknowledge FCT, FEDER, POCTI, POSI, POCI, POSC and US National Science Foundations (Grants #ECCS 0348221 and # EFRI 0836017) for their support to R&D Projects and GECAD Unit. We also thank GECAD researchers, especially Bruno Canizes, João Soares and Tiago Sousa for their help in the simulations undertaken for the illustrative case studies.

References

Amin S and Wollenberg B (2005) "Toward a Smart Grid", IEEE Power and Energy Magazine, Vol. 3, No. 5, pp. 34-41, September-October

Azevedo F, Vale Z and Oliveira P (2007) "A Decision-Support System Based on Particle Swarm Optimization for Multiperiod Hedging in Electricity Markets," IEEE Transactions on Power Systems, Volume 22, Issue 3, Pages 995-1003, August

Chaturvedi K, Pandit M and Srivastava L (2008) "Self-organizing hierarchical particle swarm optimization for nonconvex economic dispatch," IEEE Transactions on Power Systems, vol. 23, Issue: 3 Pages: 1079-1087, Aug

Energy Independence and Security Act (2007) One Hundred Tenth Congress of the United States of America, http://frwebgate.access.gpo.gov/cgi – bin/getdoc.cgi? dbname = 110_cong_bills&*docid* = *f*: *h6enr.txt.pdf* (accessed in May 2010)

Engelbrecht A, Computational Intelligence: An Introduction, John Wiley & Sons, Ltd, England, 2007, ISBN 978-0-470-03561-0

European Communities (2008) "European Energy and Transport – Trends to 2030 – update 2007," ISBN 978-92-79-07620-6, April

Giri J, Sun D and Avila-Rosales R (2009) "Wanted: A more intelligent grid," IEEE Power and Energy Magazine, Volume 7, Issue 2, Pages 34-40, March-April

Grant L, Venayagamoorthy G, Krost G and Bakare G (2008) "Swarm Intelligence and Evolutionary Approaches for Reactive Power and Voltage Control", IEEE Swarm Intelligence Symposium, St. Louis, MO, USA, September 21-23

Huang C and Wang F (2007) "An RBF network with OLS and EPSO algorithms for real-time power dispatch," IEEE Transactions on Power Systems, vol. 22 Issue: 1 Pp.96-104, Feb.

Hutson C, Venayagamoorthy G and Corzine K (2008) "Intelligent Scheduling of Hybrid and Electric Vehicle Storage Capacity in a Parking Lot for Profit Maximization in Grid Power Transactions", IEEE Energy 2030, Atlanta, GA, USA, Nov. 17-18

IEEE 14-bus (2010) test system data. Available online: http://www.ee.washington.edu/research/pstca/pf14/pg_tca14bus.htm (accessed in May 2010)

Ilic M, Black MK, Prica M (2007) "Distributed electric power systems of the future: Institutional and technological drivers for near-optimal performance", Electric Power Systems Research, Volume 77, Issue 9, Pages 1160-1177, Elsevier, July

Kodsi S and Canizares C (2003) "Modelling and Simulation of IEEE 14 Bus System with FACTS Controllers," University of Waterloo, E&CE Department, Tech. Rep. available online: http://www.power.uwaterloo.ca.

Lee K (Editor) and El-Sharkawi M (Editor) (2008) "Modern Heuristic Optimization Techniques: Theory and Applications to Power Systems," IEEE Press Series on Power Engineering, Wiley-IEEE Press, March

Lee T (2007) "Optimal spinning reserve for a wind-thermal power system using EIPSO," IEEE Transactions on Power Systems, vol. 22 Issue: 4 Pp.1612-1621, Nov

Mendes R (2004) Population Topologies and Their Influence in Particle Swarm Performance, PhThesis, Minho University, Portugal

Miller R and Malinowski J (1994) Power system operation, 3rd ed., New York; London: McGraw-Hill

Miranda V, Keko H and Duque A (2008) Stochastic Star Communication Topology in Evolutionary Particle Swarms (EPSO)IJC1IR - International Journal of Computational Intelligence Research, Vol.4 No. 2, pp.105-116

Morais H, Kadar P, Faria P, Vale Z and Khodr H (2010) "Optimal scheduling of a renewable micro-grid in an isolated load area using mixed-integer linear programming," Renewable Energy, Volume 35, Issue 1, Pages 151-156, January

Papalexopoulos, Singh H (2002)"On the Various Design Options for Ancillary Services Markets", 34th Annual Hawaii International Conference on System Sciences, Maui, Hawaii, pp. 798-805, Jan

Pereira A, Vale Z, Moura A and Pinto J (2004) "Provision and Costs of Ancillary Services in a Restructured Electricity Market", International Conference on Renewable Energy and Power Quality (ICREPQ'04), 31 March-02 April

Praça I, Ramos C, Vale Z and Cordeiro M (2003) "MASCEM: a multiagent system that simulates competitive electricity markets," IEEE Intelligent Systems, Volume 18, Issue 6, Pages 54-60, November-December

Qiao W, Venayagamoorthy G and Harley (2008) "Optimal Wide-Area Monitoring and Non-Linear Adaptive Coordinating Control of a Power System with Wind Farm Integration and Multiple FACTS Devices", Neural Networks, Vol. 21, Issues 2-3, pp. 466-475, March/April

Rahman S (2009) "Smart grid expectations [In My View]," IEEE Power and Energy Magazine, Volume 7, Issue 5, Pages 88, 84-85, September-October

Ray S and Venayagamoorthy G (2008) "Real-Time Implementation of a Measurement based Adaptive Wide Area Control System Considering Communication Delays", IET Proceedings on Generation, Transmission and Distribution, Vol. 2, Issue 1, pp. 62 - 70, Jan

Saber A and Venayagamoorthy G (2010) "Intelligent unit commitment with vehicle-to-grid -- A cost-emission optimization," Journal of Power Sources, Elsevier, Volume 195, Issue 3, Pages 898-911, February

Shahidehpour M, Yamin H and Li Z (2002) Market operations in electric power systems : forecasting, scheduling, and risk management, Institute of Electrical and Electronics Engineers, Wiley-Interscience, New York

Tai H and hÓgáin E (2009) "Behind the buzz [In My View]," IEEE Power and Energy Magazine, Volume 7, Issue 2, Pages 96, 88-92, March-April

U.S. Department of Energy (US-DoE) and the National Energy Technology Laboratory (NETL) (2008) Systems View of the Modern Grid, January

US Department of Energy (US-DoE) (2009a) "Keeping the Lights On in a New World," Report by Electricity Advisory Committee, January

US Department of Energy Electricity (US-DoE) Advisory Committee (2009b) "Smart grid system report," Report by Electricity Advisory Committee, July

Vale Z (2009) "Intelligent Power System," Wiley Encyclopedia of Computer Science and Engineering, 5 Volume Set, ISBN 978-0-471-38393-2, Benjamin W. Wah (Editor), Volume 3, Pages 1604-1613, Hoboken, NJ, January

Vale Z, Morais H, Silva M and Ramos C (2009) "Towards a future SCADA," 2009 IEEE Power and Energy Society General Meeting General Meeting, Calgary, Alberta, Canada, 26-30 July

Vale Z, Ramos C, Faria P, Soares J, Canizes B and Khodr H (2009) "Ancillary Service Market Simulation", IEEE T&D Asia, Seoul, Korea, October

Vale Z, Ramos C, Faria P, Soares J, Canizes B and Khodr H (2010) Ancillary Services Dispatch using Linear Programming and Genetic Algorithm approaches, The 15th IEEE Mediterranean Electrotechnical Conference (MELECON 2010), La Valletta, Malta, 25 - 28 April

Vale Z, Ramos C, Morais H, Faria P and Silva M (2009) "The role of demand response in future power systems," IEEE - T&D Asia 2009, Seoul, Korea, 27 – 30 October

Venayagamoorthy G (2009) "A Successful Interdisciplinary Course on Computational Intelligence", IEEE Computational Intelligence Magazine – A special issue on Education, Vol. 4, No. 1, pp. 14-23, February

Venayagamoorthy G (2009) "Potentials and Promises of Computational Intelligence for Smart Grids," Proceedings of 2009 IEEE PES General Meeting, Smart Grid Panel, Calgary, Canada, July

Wu J, Zhu J, Chen G, et al. (2008) "A Hybrid Method for Optimal Scheduling of Short-Term Electric Power Generation of Cascaded Hydroelectric Plants Based on Particle Swarm Optimization and Chance-Constrained Programming," IEEE Transactions on Power Systems, vol. 23 Issue: 4 Pp. 1570-1579, Nov

Zimmerman R and Gan D (1997) "Matpower Manual", USA: PSERC Cornell Univ